Responsible
Science or
Technomadness?

CLONING

Edited by
Michael Ruse & Aryne Sheppard

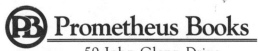

PB **Prometheus Books**

59 John Glenn Drive
Amherst, New York 14228-2197

Published 2001 by Prometheus Books

Inquiries should be addressed to
Prometheus Books
59 John Glenn Drive
Amherst, New York 14228–2197
VOICE: 716–691–0133, ext. 207
FAX: 716–564–2711
WWW.PROMETHEUSBOOKS.COM

05 04 03 02 01   5 4 3 2 1

Library of Congress Cataloging-in-Publication Data

Cloning : responsible science or technomadness? / edited by Michael Ruse and Aryne
    Sheppard.
        p. cm.
    Includes bibliographical references.
    ISBN 1–57392–836–4 (pbk. : alk. paper)
    1. Cloning—Social aspects. I. Ruse, Michael. II. Sheppard, Aryne.
III. Contemporary issues (Amherst, N.Y.)

QH442.2 .C5678 2000
174'.966—dc21   00–045722

Printed in the United States of America on acid-free paper

Contents

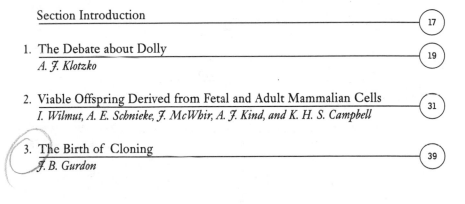

5

CLONING

II. Bioengineering: Animals and Plants

III. Human Cloning: The Questions of Dignity and Identity

IV. HUMAN CLONING: THE CASE AGAINST

V. HUMAN CLONING: THE CASE FOR

VI. HUMAN CLONING: SOCIETAL QUESTIONS

VII. Human Cloning: Medical Implications

VIII. Religion: Official Statements

IX. Religion: Perspectives

X. Policy and Regulation

◯— Introduction

In 1932 the English writer Aldous Huxley published his best-known work, a futuristic novel: *Brave New World*. As the grandson of "Darwin's bulldog," the eminent Victorian morphologist and essayist Thomas Henry Huxley, and as the younger brother of Britain's then most famous, living zoologist Julian Huxley, Aldous Huxley was certainly *au courant* not only with the science of his day but also with speculations and hypotheses about the possible directions of future science and technology. This knowledge he used to the full, letting his creative imagination range freely and fertilely over the territory, as he sketched out a vision of the future: a society which horrifies even today in its absurd acceptance of and frenzied delight in soul-destroying ideas and practices fueled by advances in empirical knowledge. And nothing quite chills and titillates the reader—then and now—than the production of made-to-order humans, designed to fit exactly within the preordained roles demanded by that society for efficient functioning.

Nearly two and a half thousand years before, in his dialogue the *Republic*, the Greek philosopher Plato had speculated about an ideal state which would have three orders of citizens. The men and women of gold who ruled; the men and women of silver who ran and defended the state; and the men and women of iron or brass who were the regular citizens, producing and distributing the goods a state needs for daily life. In a wild parody of this, Aldous Huxley supposed that his future society would turn out Alphas at the top, needed to do the very highest form of thinking and calculating, and then go down the Greek alphabet to the Deltas and the Epsilons, who are little more than morons but perfect for the mind-destroying jobs that must be performed at the production belt of a modern factory. (Remember that Huxley was writing during the heyday of Henry Ford and his mass-produced automobiles.)

What made Huxley's vision truly dreadful and threatening was the fact that these ready-made, off-the-shelf humans came—like one of Ford's Tin Lizzies—with a built-in and specifiable uniformity. You need a Delta to do some boring, repetitive job? Why not have three or four of them while you are at it? Huxley's new world had mastered the technology of producing viable organisms from the cells of others, of "cloning." Thus the scientists and technicians were able to produce multiple copies of the same human. Having in mind precisely the sort of person wanted, one could then take out the element of unpredictability and set about producing them wholesale. Efficiency may have been produced certainly, but surely at the cost of just about everything that we think makes it worthwhile and important to be a human being. One's individuality, one's worth as an independent person, is gone forever. One is no more than a copy, a print from the press, an automobile which emerges from the factory absolutely identical to millions of others which went before and which come after. One is no longer a unique being. One is a machine-produced artifact, with no more standing than any other such artifact.

When *Brave New World* appeared, any thoughts that human cloning might truly occur were away in the realms of fiction or the distant future. But now the future is here, or almost so. In 1997 Scottish researchers announced that they had managed to clone an adult mammal, a sheep. The world's most famous lamb, Dolly, was born from a cell taken from her mother's udder. A sheep of course is not a human, but cloned monkeys followed closely on Dolly, and one suspects that humans cannot be far behind, if indeed they have not already been produced in some unannounced experiment somewhere in the world. Huxley's fantasy is uncomfortably close to reality, or at least to possibility. Hence, as we move into the new millennium, one of the things we must tackle is the question of the desirability or otherwise of cloning—of animals and plants, and most particularly of humans. We must look at the technology which lies behind the procedure, we must ask about the benefits which might accrue for us all (for instance from the production of desirable foodstuffs or the easy access to medically useful substances), and above all we must ask whether human cloning can ever be a permissible or desirable practice. And if we decide that a positive case can be made for human cloning, we must ask how positive that case may be. Do we need to make distinctions between cloning for experimentation and discovery, cloning for spare parts, and cloning for producing new humans? Should cloning be a matter of personal choice, without state interference or regulation? Or should it be allowed if at all only under very strictly regulated circumstances? Or does the right course of action lie somewhere in the middle?

Already, Dolly has generated much interest and discussion, by the general public, by the biological and medical professions, by professional ethicists and others working on social policies, by theologians and others with religious commitments and interests, and by governments. There is much interest and discussion and disagreement. Not so much about the technology, which seems to be going ahead splendidly. But rather about the implications. Some think that cloning is going to be a cornucopia, mass-producing newly designed plants and animals, in a way hitherto unsupposed. Others are not so sure, and see roadblocks both material and social. Some worry—as Aldous Huxley intended us to worry—that if humans are cloned, they will lose their identity, their true worth as human beings. Others think these fears are much exaggerated. Are identical twins not fully worthwhile human beings? Some think that any technology like cloning is an unwarranted interference in the natural reproductive processes and as such must be judged wrong. Others see cloning for humans as being on a par with a face-lift—not to be done by amateurs but with no deep moral or religious significance. Thus, they see it as something which ultimately should be a matter of complete, personal choice. And so the debate goes on.

We editors think that this is an important debate and it is one which should take place. We think also that it is a debate where a fairly detailed understanding of the science and of the implications is needed, especially if meaningful conclusions are to be drawn. For this reason, we have put together this collection, including discussions of the science, of the practical implications, of the moral and social and religious implications for human beings, and of governmental and other reactions. It is a cutting-edge discussion and we have striven to make our collection as up-to-date as possible. We hope that you will find the collection useful and that, with its aid, you can draw some conclusions about this very significant technological development. We have tried to present opposing views where pertinent, realizing that blunt disagreement is the nature of discussions such as these and that if one is to draw meaningful personal conclusions, one must do so not only on the strengths of the arguments of the side you favor but also in the light of criticisms made by those whose sides you ultimately reject. Above all else, we hope that you find this to be a debate which is as intrinsically interesting as it is important.

CLONING

Because already the literature has grown to a very large size, we had a large pool from which to draw. We have added a "further reading" discussion to guide you to material which we simply could not include, and also to help you keep abreast of developments in the future. We recognize that some discussions of cloning tend to get a bit complex, with a high jargon factor. This is by no means true of all or even most pieces on the subject, but it is true of some. However, we did not want to dumb down our selection, choosing only those articles which were written for people who have no sophisticated training in any area of human inquiry. Some pieces—the original announcement of Dolly for instance—simply had to be included, even though for the nonexpert they may be hard going in places. But tempering the wind to the shorn lamb—to use an appropriate metaphor and a dreadful pun—we have added a glossary of technical terms, and we think this will enable you to follow even the more difficult discussions.

There is one more thing to do before we can point you forward and ask that you to get stuck into the topic at issue. You are going to find that, especially when we get to discussions of human cloning, many writers are presupposing some knowledge of moral philosophy, or at least are working within the context of a moral theory. There is nothing which should be impossibly difficult, even for those with little or no philosophical background. It may, however, help your comprehension if we say just a few clarifying words, starting with the fact that moral philosophy is concerned with questions about what we should do and why we should do what we should do. Many religions have a moral dimension. For instance, Christianity has at its heart the love commandment: the demand that you should love your neighbor as yourself. There are also secular moral philosophies, that is to say systems which make no direct reference to a deity (although as a matter of fact many who have articulated such philosophies have been deeply religious people).

In modern thought there are two main such secular moral philosophies. The first is known as *utilitarianism*: it is a "consequentialist" theory, judging thoughts and actions by their results, and usually measures moral success in the happiness achieved or the unhappiness avoided. There has been much debate about precisely what one means by happiness: the English moral philosopher John Stuart Mill declared that it was better to be Socrates dissatisfied than a pig satisfied. Aldous Huxley's *Brave New World* is based on what many would regard as a bastardized form of utilitarianism: the aim of his society is predicated on and solely on a state of nigh-mindless contentment, with its citizens like cows chewing the cud rather than Socrates arguing about the meaning of life.

The great alternative to utilitarianism comes from the mind of an eighteenth-century German philosopher: *Immanuel Kant*. He insisted that it is the good will which is important rather than mere consequences (the technical term for such a position is "deontology"). More than this, he argued that we must always obey some supreme rule of conduct, something which Kant called the "Categorical Imperative." This came in various versions, but the one of most concern to us is that which urged us to treat human beings as ends in themselves rather than as simply means to an end. To show what Kant means by this, consider as an example the horrendous parking problems in downtown New York, with people double-parking and not paying their fines and so forth. Suppose the authorities decided on a few public executions for miscreants. No doubt there would be a rapid improvement in the situation. Perhaps a utilitarian (of a peculiarly sadistic kind or who always walks to work!) might even judge this a good act, for the overall, end happiness would be much improved. But the Kantian would always judge it wrong because the unfortunate victims would be punished way beyond the demerits of their case, simply to make an example for the benefit of others. They would be means not ends in themselves.

Clearly the Kantian makes the status of the human—the dignity or worth of each and every one of us—absolutely central. As you can imagine, this has been a claim that has much attracted the attention of those writing about the moral status of human cloning, although as you will see not all would interpret the relevance of Kant—who of course himself had absolutely no idea whatsoever about cloning—in precisely the same way. But how people do interpret Kant and what you think of their arguments will be for you to find out and to decide as you read the following pages, a task which we now invite you to begin.

DOLLY

The birth of lamb number 6LL3, better known as "Dolly," was announced to the world early in 1997. She was the product of arduous and skilled work by Dr. Ian Wilmut and his team at an institute near Edinburgh, in Scotland. What made her remarkable was that she was not produced in the normal natural way, but came rather directly from one of the mammary cells taken from her mother. She was a "clone"—with a nucleus genetically identical to her mother and with no father in sight or needed at all. To say that this was an exciting event is to underestimate the coverage in the popular press both in Britain and in North America. The animal's birth was treated with the hoopla (and sometimes wild speculation) that one usually finds reserved for sightings of Elvis as reported in the supermarket tabloids. But not just at this popular level, for almost immediately serious thinkers and makers of social policy were involved. In less than a month, President Bill Clinton was issuing executive orders banning federal funds for attempts to produce human clones.

The readings in this section introduce you to the whole issue, setting it in context, and presenting Dolly herself. Arlene Judith Klotzko gives a brief history of the actual Dolly event, and about the reactions in the press and then at more official levels. It is clear that this all became as much a media as a scientific event, but no less significant or important for all that. Next we have the seminal article by Ian Wilmut and his associates, telling of the work which was needed to produce a cloned sheep. Note that Dolly was unique in coming from the cell of an adult animal, but that there were other lambs produced from the cells of embryonic sheep. At one level you might think that, although this may be scientifically significant, at the moral or social or religious level whether a clone comes from an embryo or an adult is not so important. Perhaps so, although do note also that if one can clone adults, then this opens whole new possible dimensions of action: most obviously, one can wait until one knows the nature of the adult before one clones. One has that much more control over the product.

Finally in this section, we offer a piece by J. B. Gordon. This deals with some of the historical background to cloning, showing how Dolly is the end result of much pertinent earlier work on the cell and specifically on manipulation of cellular parts, making them do things that do not (or only rarely) occur in nature. Scientifically and technically, producing Dolly was a wonderful achievement, but it was not something which came entirely out of the blue. There was a long history. And indeed, one suspects that Dolly herself will become part of that history, as technology gets ever-more sophisticated. There are now reports that Dolly ages more rapidly than normal sheep. This may be true, but one doubts that this will be the last limiting word or finding. Before long, the practice of cloning will be so advanced that the Dolly event will look like those midcentury, room-filling computers as compared to today's laptop.

1 The Debate about Dolly
Arlene Judith Klotzko

THE SCIENTIFIC BACKGROUND

On February 23, 1997, a British newspaper, *The Observer*, broke a story that truly shocked the world. Dr. Ian Wilmut and his team at the Roslin Institute, near Edinburgh, Scotland, had succeeded in a task many had believed to be impossible. They had cloned a mammal—a sheep, named Dolly—from an adult mammary cell. (In the same experiment, the scientists also cloned sheep from cell lines composed of fetal and embryo cells.)[1]

In order to create Dolly, the scientists performed nuclear transfer. Two hundred and seventy seven times. Again and again, they removed a nucleus from an egg cell and replaced that nucleus with one that had been taken out of a mammary cell from an adult sheep. They applied an electric current, which caused the egg and its new nucleus to fuse and develop into an embryo. Any embryos that resulted were implanted into surrogates. Dolly was the only lamb to be born.

Reprinted by permission from *Bioethics* 11 (1997): 427–38.© Blackwell Publishers Ltd. 1997.

CLONING

She is virtually a genetically identical copy of the adult sheep whose mammary cell was used in the experiment—hence, a clone. Dolly can be understood as a kind of later-born identical twin but, unlike those twins that occur naturally, there is some genetic difference between Dolly and her twin. Although the egg with which the nucleus of the adult sheep cell was fused had been stripped of its nucleus, it was not empty; it contained its own DNA—mitochondrial DNA.

Before this experiment, conventional wisdom held that a mammal could not be cloned from an adult cell. Unlike early embryo cells, which are totipotent—capable of becoming any and every cell in the body—adult cells are differentiated. In such cells—those that form the skin, muscle, and brain, for example—genes not needed to perform the required specialized function are switched off. In contrast, an undifferentiated cell can give rise to any cell in the body because it is capable of activating any gene on any chromosome.

In essence, the Roslin team had to trick the adult cell's DNA into reverting to its undifferentiated past. And they did—by placing the mammary cells in a culture and starving them of nutrients for several days. In this quiescent state, it is believed that few if any genes remained switched on. When the nuclei were removed from the adult cells, placed next to the enucleated egg cells, and fused by electricity, the eggs were able to reprogram the donor nuclei into behaving as if they had come from undifferentiated cells. The precise mechanism of this reprogramming is still not completely understood.

Cloning of sheep and cows had been done successfully for some time, but only from early embryonic cells, and only in relatively small numbers at each attempt. In March 1996—almost exactly one year before the world got news of Dolly—the Roslin team announced a stunning breakthrough—the birth of Megan and Morag, two sheep cloned from mature embryo cells.[2]

The technique that they used—nuclear transfer performed with cells that had grown in a culture and then been rendered quiescent—was the same technique used to clone Dolly. Five identical lambs were born; two died at birth, and a third soon after. Both Megan and Morag are now pregnant—one due to give birth in May and the other in July. From one fertilized egg, Wilmut and his colleagues were able to grow a collection of cells in whatever quantities they wished, and thereby enable the production of very large numbers of genetically identical animals—even entire flocks.

The ultimate purpose of these experiments is to produce—efficiently and predictably—large numbers of transgenic animals (animals with human genes) that can secrete pharmaceutical compounds in their milk. In Britain, the creation of Megan and Morag was greeted with frenzied media speculation about creating legions of identical copies of animals. And humans. Concern was somewhat muted, however, because they were clones of embryos—not existing adults. Perhaps, for that reason, the American media did not take note.

THE MEDIA RESPONSE

In contrast, the cloning of Dolly produced an explosion of coverage in the American print and television media. Cloning stories appeared on the covers of *Time* and *Newsweek*. Most newspapers ran front-page stories, and the coverage persisted for weeks. Many stories appeared accompanied by diagrams of nuclear transfer.

Public consternation, already at fever pitch, only grew when, on March 2, it was announced that two monkeys had been cloned from embryo cells at the Primate Research Center in Beaverton, Oregon. Although this experiment marked the first time that live primates have been produced through cloning, embryo cells were used—not adult cells. The public inferred—erroneously—that we were now a step closer to cloning humans. As more than one observer dryly noted, last week it was a sheep, this week a monkey, so next week it will probably be a human.

There has been excellent reporting on the scientific accomplishments of Ian Wilmut—notably in the *New York Times* and the *Washington Post*. But there has been a lot of irresponsible reporting as well. *Newsweek*, for example, wondered whether society could "stuff the cloning genie back into the bottle," and compared Wilmut's discovery with nuclear bombs (Hiroshima and Nagasaki) and chemical weapons (the attack on the Tokyo subway).[3]

Americans were treated to descriptions of Mary Shelley's *Frankenstein*, armies of drones, and clone farms to produce spare parts. One particularly imaginative journalist painted an image of grandiose dictators cloning generations of themselves instead of building monuments[4]—evoking thoughts of another Shelley—Percy Bysshe. His poem, "Ozymandias," describes a statue of a once powerful ruler, reduced by the passage of time to a pedestal, a shattered visage, and two trunkless legs. Cloning could have provided the story of Ozymandias with a happy ending. Or, perhaps, no ending.

The sensational character of U.S. press coverage was matched in the UK, where the most egregious example came from the *Daily Mail*. The headline of its February 24 issue asked, "Could We Now Raise the Dead?" obviously confusing replication with resurrection. George Annas has written that "novels such as *Frankenstein* and *Brave New World* and films such as *Jurassic Park* and *Blade Runner* have prepared the public to discuss deep ethical issues in human cloning."[5] I emphatically disagree.

CLONING

THEOLOGICAL VOICES

The American media has been full of bioethics and bioethicists on talk shows, in television interviews, and in newspaper and magazine articles. While American bioethics—fortunately or unfortunately, depending upon one's point of view—often has a consequentialist cast, the cloning discussion has had a pronounced deontological flavor. To be sure, opponents of human cloning made secular arguments—consequentialist and deontological—but many of the concerns articulated had a religious character that is particularly noteworthy.

Dr. Stanley Hauwerwas, a divinity professor at Duke University, told the *New York Times* that he saw "a kind of drive behind this for us to be our own creators. In the same article, Dr. Kevin FitzGerald, a Jesuit priest and geneticist, said that a human clone would "have to have a different soul."[6] In a commentary in the *Washington Post*, theologian Nancy Duff stated her view that "if there were ever a Tower of Babel—which originally was an attempt to elevate ourselves through human accomplishments to the level of God—surely this is it."[7]

According to Daniel Callahan, American bioethics began with theologians—most notably, Joseph Fletcher and Paul Ramsey—but they were soon joined by lawyers, philosophers, and social scientists, and the field was secularized.[8] The ethical discussion of cloning, however, seems to have taken us back in time. And the customary public and media excitement over the latest advances in medical technology was eclipsed by talk of moral repugnance, evil, wrongness, playing god, and impermissible interventions. In this respect, there was a striking difference between American and British media coverage. In the UK, the overall character of the reporting was secular; religious arguments did appear, but they were few and far between.

Ronald Dworkin has noted the profound ambivalence of Americans on the subject of religion. Despite its constitutional separation of church and state, the United States is among the most religious of modern Western countries. And in the tone of its most powerful religious groups, by far the most fundamentalist.[9] In the opposition to abortion and to euthanasia, the Catholic church has played an important role. As have fundamentalist Christians—particularly in regard to abortion, fetal tissue research, and embryo research.

But there is no anticloning lobby; religious objections to human cloning have been much more broadly based than in those other contexts. And they have been influential. The National Bioethics Advisory Commission (NBAC) devoted much of its first public hearing on human cloning to taking testimony from four Protestant theologians, and their Catholic, Jewish, and Muslim counterparts.[10]

The consequentialists might have been outshone but they were not invisible. As Daniel Callahan observed, "the argument about Dolly saw two camps instantly formed—one was alarmed by the development and opposed to any further movement toward cloning humans; the other (seemingly much smaller) touted a potential gain in health and more reproductive choice if cloning went forward."[11] And, of course, consequentialist arguments can cut both ways. They can also point to the dire consequences of human cloning. And they did.

THE GOVERNMENT RESPONSE

Despite the enormous impact American bioethics has had in the United States and worldwide—Prof. Alexander Capron, a member of NBAC, has noted that "the mere mention of the term 'bioethics' stirs up controversies in some quarters—particularly among right-to-life advocates."[12] In the late 1980s, the agenda of the Biomedical Ethics Advisory Committee—created by the Congress to study issues raised by the new genetics—ran afoul of the politics of abortion. The committee expired in 1990, after holding only two meetings.

Subsequently, there was a move to set up a national bioethics commission to be chartered by Congress. When Congress failed to enact legislation, the President's Office of Science and Technology Policy proposed to charter a bioethics commission as a subcommittee of the National Science and Technology Council which advised the Assistant to the President for Science and Technology. President Clinton issued an executive order creating NBAC in October 1995.[13] The members were not named until July 1996. Unless its charter is renewed, the commission is scheduled to go out of business in October 1997—one year after holding its first meeting.

In the words of Dr. Arthur Caplan, NBAC was given a "very low priority.... All of a sudden, cloning explodes and the president looks desperately ... for help and advice. The only group he can go to is the National Bioethics Advisory Commission."[14] On February 24, the president asked NBAC for a report on the legal and ethical implications of the cloning of Dolly—a report playfully described by one British commentator as "a quick-roasted ethical attitude without mint sauce within 90 days."[15]

On March 4, 1997, President Clinton issued an executive order banning the use of federal funding for human cloning research. Such a ban is a mechanism that has been seen before. A ban on the use of federal funds for research on human fetal tissue was instituted in 1988 and lifted only in 1993. A ban on federal funding for human embryo research has been in effect since 1994. The president also asked privately funded scientists to halt any human cloning research until NBAC issues its report in late May.

But legislators were not waiting for the report. On February 27, 1997, legislation (S368) was introduced in the Senate by Sen. Christopher Bond;

(a) IN GENERAL—No federal funds may be used for research with respect to the cloning of a human individual.

(b) DEFINITION—For purposes of this section, the term "cloning" means the replication of a human individual by the taking of a cell with genetic material and the cultivation of the cell through the egg, embryo, fetal, and newborn stages into a new human individual.

On March 5, 1997, legislation (HR 923) was introduced in the House of Representatives by Rep. Vernon Ehlers. His Human Cloning Prohibition Act is broader than the Senate bill. It not only denies federal funds, it makes using "a human somatic cell for the process of producing a human clone" a civil wrong, and assesses a civil penalty of up to $5,000.

As of this writing, legislators in New York, California, Illinois, Alabama, Florida, Maryland, Missouri, New Jersey, North Carolina, Oregon, South Carolina, and West Virginia have introduced bills that would do one or more of the following: prohibit human cloning research, prohibit use of state funds for such research, urge the president and Congress to do the same with federal funds, create a panel to advise the state legislature, or make human cloning a criminal offense.

In the Senate, a remarkable effort to foster an informed public (and legislative) discussion was quickly put together. On March 12, 1997, a hearing was held by the Senate Labor Committee's Subcommittee on Public Health and Safety. Dr. Ian Wilmut was one of ten witnesses invited to testify. The chair of the subcommittee, Senator William Frist, a former heart transplant surgeon, stated that he hoped to use the hearing to begin a public discussion about cloning, and that the precondition for such a discussion was a thorough understanding of the science and of the underlying facts. The entire hearing was broadcast several times on television, and Dr. Wilmut's remarks were reported in virtually every newspaper and television news program.

Before the session, Dr. Wilmut met with Senator Frist and Senator Kennedy, the subcommittee's ranking Democratic member. In an interview, he characterized that meeting as follows:

> It was clear that their aim was to allow careful thought before legislation. They were concerned that there was a knee-jerk response that said "We must stop that,"

which could inadvertently prohibit uses which society would accept. I was very impressed with the senators and their staff assistants. And I hope they achieve their objective, because one difficulty with this subject is that people use words carelessly. There may be uses of nuclear transfer with human cells that do not involve producing a new person. It is very important to make sure that in considering the prohibition of the production of new people, you don't inadvertently prohibit acceptable uses.[16]

Dr. Wilmut did not participate in the public hearings held by the National Bioethics Advisory Commission, in Washington, on March 13 and 14. When asked about his absence from these meetings, he said that he had not been invited to attend:

I am surprised that my colleagues and I have not been asked to present any information to the National Bioethics Advisory Commission, either in person or in writing.

In an April 22 speech before the National Press Club, Dr. Neal Lane, Director of National Science Foundation, said that the newfound ability to clone a mature sheep demands extensive public discussion and debate. But he expressed concern at the low level of scientific literacy in the United States. Senator Frist addressed this deficit by asking the two scientist witnesses—Dr. Harold Varmus, Director of the National Institutes of Health, and Dr. Wilmut—to explain the science of cloning and its implications in a way that the public could understand. And they did—brilliantly.

But, after the saturation coverage of the Frist hearing, the media seemed to lose interest. The two-day NBAC hearing that began the next day received meager coverage, and their subsequent meetings even less. There will be a burst of media interest when their report is unveiled. And when the ban on federal funding of human cloning research expires, there will be—one hopes—a public discussion about what should be done in the face of the unusual, if not unique, American legal situation concerning assisted reproduction.

THE LEGAL LANDSCAPE

In the United States, there is no comprehensive regulatory scheme for IVF and embryo research. Human IVF has been developed and advanced in the private sector. More than 400 clinics operate with no federal money and little federal oversight. Guidelines have been developed by the industry, but compliance is voluntary.[17] There is also a set of ethical considerations, published by the American Society for Reproductive Medicine in 1994.[18]

According to biologist Prof. Lee Silver, "there are hundreds or private IVF clinics in America, where doctors and technicians are capable of... artificially inseminating donated human eggs. Cloning would be no problem to such people."[19] There is no existing federal prohibition other than the president's temporary one—involving the cloning of humans or human embryos. And no federal laws regulating embryo research. The subject is such a contentious one that Congress has simply been unable to address it.

In 1993, following a report that scientists at George Washington University, in Washington, D.C., had cloned human embryos, by embryo splitting, [20] there was a blast of media coverage and public concern somewhat similar to recent events. But the American scientists had not employed nuclear transfer; they used a less advanced technique, called blastomere separation, by which the totipotent cells in two to eight cell embryos are made to separate, and then form smaller than normal embryos. And they used private—not federal—research funds.[21]

In February 1994 a Human Embryo Research Panel was appointed by the Director of the National Institutes of Health. Later that year, the panel recommended that federal funding be allowed for certain types of embryo research. Included within this category was research on those embryos remaining after IVF or preimplantation diagnosis, up to the appearance of the primitive streak (day 14), and on embryos purposely created when a "compelling case" can be made for the scientific and therapeutic value of the research. Among the categories of research determined to be unacceptable for federal funding was research involving cloning, followed by transfer.[22]

When the report was released, the political uproar was immediate and fierce. In December 1994 President Clinton banned the use of federal funds to create embryos for research purposes. Congress went even further—extending the ban to research on so-called spare embryos. In amendments to the appropriations bills funding the National Institutes of Health in 1996 and 1997, Congress prohibited the NIH from using federal funds to finance any research involving the destruction of embryos.

The Legal Outlook

The antiabortion lobby in the United States is extremely powerful. While President Clinton did lift the ban on fetal tissue research funding, it seems unlikely that he would back a repeal of the embryo research ban now in place. And even if he did, Congress—with a Republican majority in both houses—would almost certainly not favor repeal.

At the Senate hearing, Drs. Varmus and Wilmut argued strongly for the preservation of a distinction between research to achieve the cloning of human beings and research on human cloning at the cellular level:

> One potential use of [the latter] ... would involve taking differentiated cells, such as skin cells, from a human patient suffering a genetic disease. These cells inevitably have a limited potential to do other things. One could take these cells—and using the technique that we have developed—get them back to the beginning of their lives by nuclear transfer into an oocyte to produce a new embryo.
>
> From the new embryo, you would be able to obtain relatively simple, undifferentiated cells, which would retain the ability to colonize the tissues of the patient. Once these cells were in the laboratory, there would be the ability to make genetic changes or even add a gene.

Thus, for human cloning at the cellular level to achieve its medical promise, it will become necessary to do research on very early embryos, created specifically for this purpose. If the ban on federal funding for embryo research remains, such research could be crippled.

There would certainly be sufficient votes in Congress to attach to the NIH appropriations bill a human cloning research ban that is similar to the existing one concerning embryo research. In this public and political climate, that probably will not be enough. Consensus is not a by-product of the adversarial American legal system, and controversies about issues such as abortion and assisted suicide are often articulated as contests between conflicting rights. But there is remarkable consensus, among the public, the politicians, and most bioethicists, that cloning of human beings—either because of its nature or its results—is too problematic to be allowed.

So there will be great pressure to legislate. Perhaps the best argument in favor of a legislative ban—as opposed to the extension of the federal funding ban now in place—is that a statute could prohibit privately funded research. The disadvantages of such an option are that it would be difficult to undo. A statute will surely wend its way through the process. And it would be difficult to imagine President Clinton vetoing such a statute. If the consequentialists—and the scientists—can be heard, the legislation will be narrowly drafted to ban research leading to the cloning of human beings, and it will be silent on—or, even better, it will regulate—cloning at the cellular level.

Politics and Ethics

Logic would dictate that a long overdue comprehensive scheme to regulate IVF and embryo research would be designed—and cloning addressed within that framework. But, unfortunately, when the discussion turns to reproductive questions in the United States, the imperatives are more often political than logical. And, of course, there are issues, like the status of the embryo, that provoke profound moral disagreement—disagreement that remains impervious to someone else's logic. The UK's Human Fertilisation and Embryology Act is a legislative tour de force in that it creates a legal framework for embryo research in the absence of moral consensus.[23]

But it must be remembered that questions of abortion (and embryo research) are far less contentious in the UK than in the United States, where, as Daniel Callahan has noted, bioethical questions often get "overlaid with ideology and swept up in the culture wars."[24] Particularly when they are related to abortion. NBAC "has explicitly not been asked to address issues of embryo research and abortion, ethical problems that are ... so contentious that any committee venturing into those waters would likely sink without a bubble."[25] Because human cloning at the cellular level would involve the deliberate creation of embryos not destined to be implanted, such research does not appear to have a particularly bright future in the United States.

Those who try to navigate waters that are both morally troubled and highly political may find the experience enormously frustrating. Prof. Alta Charo, a member of NBAC, who also served on the Human Embryo Research Panel, told the *New York Times* that the panel's work had taught her a valuable lesson:

> That group, she said, relied on logic to make its case that research with early human embryos was ethically acceptable. "The logic was airtight, but it did not change anybody's mind and there was a lot of resentment," Ms. Charo said. Now, she said, she realizes that "logical arguments are only rationalizations for gut feelings or religious viewpoints." And, she said, that is where the group ought to start in analyzing what it wants to say about the cloning of humans.[26]

On January 22, 1993, when President Clinton announced the lifting of the ban on federal funding for fetal tissue research, he said that "we must free science and medicine from the grasp of politics." Surely, we must, but all too often we don't. Let us hope that bioethics—and with it, the promise of human cloning at the cellular level—will not suffer a similar fate. [27]

NOTES

1. I. Wilmut, A. E. Schnieke, J. McWhir, A. J. Kind, and K. H. S. Campbell, "Viable Offspring Derived from Fetal and Adult Mammalian Cells," *Nature* 385 (February 27, 1997): 810–13.

2. K. H. S. Campbell, J. McWhir, W. A. Ritchie, and I. Wilmut, "Sheep Cloned by Nuclear Transfer from a Cultured Cell Line," *Nature* 380 (March 7, 1996): 64–66.

3. S. Begley, "Little Lamb Who Made Thee?" *Newsweek* 56 (March 10, 1997).

4. J. Kluger, "Will We Follow the Sheep?" *Time* 71 (March 10, 1997).

5. G. Annas, "Human Cloning: Should the United States Legislate Against It?" *ABA Journal* 80 (May 1997).

6. G. Kolata, "With the Cloning of a Sheep, the Ethical Ground Shifts," *New York Times*, February 24, 1997, p. A1.

7. N. Duff, "Clone with Caution: Don't Take Playing God Lightly," *Washington Post*, March 2, 1997, p. Cl.

8. D. Callahan, "Bioethics and the Culture Wars," *The Nation* 24 (April 14, 1997).

9. R. Dworkin, *Life's Dominion: An Argument about Abortion, Euthanasia and Individual Freedom* (New York: Knopf, 1993), p. 6.

10. L. Witham, "Christians Oppose Human Clones," *Washington Times*, March 14, 1997, p. 3.

11. Callahan, "Bioethics and the Culture Wars," p. 23.

12. A. M. Capron, "An Egg Takes Flight: The Once and Future Life of the National Bioethics Advisory Commission," *Kennedy Institute of Ethics* 7, no. 1 (1997): 67.

13. Ibid., p. 65.

14. G. Kolata, "Little-Known Panel Challenged to Make Quick Cloning Study," *New York Times*, March 18, 1997, p. C9.

15. T. Radford, "Well, Hello Dolly…," *Guardian*, February 27, 1997, p. 5.

16. The comments of Dr. Ian Wilmut, quoted here and elsewhere in this article, were derived from a wide-ranging interview that will be published early next year—as part of a special section on cloning—in the *Cambridge Quarterly of Healthcare Ethics*.

17. Congressional Research Service, *CRS Report for Congress*, 97–335 SPR (March 11, 1997).

18. Ethics Committee of the American Fertility Society, "Ethical Considerations of Assisted Reproductive Technologies," *Fertility and Sterility Supplement* 62 (November 1994).

19. McKie, "But Will There Ever Be Another You?" *The Observer* 18 (March 2, 1997).

20. J. L. Hall, D. Engel, P. R. Gindoff, et al., "Experimental Cloning of Human Polyploid Embryos Using an Artificial Zona Pellucida," The American Fertility Society conjointly with the Canadian Fertility and Andrology Society, Program Supplement, Abstracts of the Scientific Oral and Poster Sessions, Abstract 0-001, S1 (1993).

21. See note 17.

22. "What Research? Which Embryos?" *Hastings Center Report* 25, no. 1 (1995): 36.

23. A. J. Klotzko, "The Regulation of Embryo Research under the Human Fertiliza-

CLONING

tion and Embryology Act of 1990," in *Conceiving the Embryo,* ed. D. Evans (The Hague: Martinus Nijhoff Publishers, 1996), pp. 303–14.

24. Callahan, "Bioethics and the Culture Wars," p. 23.

25. J. Palca, "New National Bioethics Commission—Maybe," *Hastings Center Report* 26, no. 1 (January–February 1996): 5.

26. See note 14.

27. I am grateful to the following persons for their varying contributions to this article: Dr. Kenneth Boyd; Mrs. Patricia Boyd; Dr. Tony Hope; Ms. Susan Joffe; Ms. Rachel McWilliams; Ms. Laurie Petrick; Dr. Zbigniew Szawarski; Dr. Ian Wilmut, the members of his research team, and his personal assistant, Ms. Jaki Young. And my special thanks to Ms. Elizabeth Graham, Senior Information Officer, The Wellcome Centre for Medical Science, for her indefatigable research support and unfailing good humor.

2 Viable Offspring Derived from Fetal and Adult Mammalian Cells

I. Wilmut, A. E. Schnieke, J. McWhir, A. J. Kind, and K. H. S. Campbell

Fertilization of mammalian eggs is followed by successive cell divisions and progressive differentiation, first into the early embryo and subsequently into all of the cell types that make up the adult animal. Transfer of a single nucleus at a specific stage of development, to an enucleated unfertilized egg, provided an opportunity to investigate whether cellular differentiation to that stage involved irreversible genetic modification. The first offspring to develop from a differentiated cell were born after nuclear transfer from an embryo-derived cell line that had been induced to become quiescent.[1] Using the same procedure, we now report the birth of live lambs from three new cell populations established from adult mammary gland, fetus and embryo. The fact that a lamb was derived from an adult cell confirms that differentiation of that cell did not involve the irreversible modification of genetic material required for development to term. The birth of lambs from differentiated fetal and adult cells also reinforces previous speculation[1,2] that by inducing donor cells to become quiescent it will be possible to obtain normal development from a wide variety of differentiated cells.

CLONING

It has long been known that in amphibians, nuclei transferred from adult keratinocytes established in culture support development to the juvenile, tadpole stage.[3] Although this involves differentiation into complex tissues and organs, no development to the adult stage was reported, leaving open the question of whether a differentiated adult nucleus can be fully reprogrammed. Previously we reported the birth of live lambs after nuclear transfer from cultured embryonic cells that had been induced into quiescence. We suggested that inducing the donor cell to exit the growth phase causes changes in chromatin structure that facilitate reprogramming of gene expression and that development would be normal if nuclei are used from a variety of differentiated donor cells in similar regimes. Here we investigate whether normal development to term is possible when donor cells derived from fetal or adult tissue are induced to exit the growth cycle and enter the G0 phase of the cell cycle before nuclear transfer.

Three new populations of cells were derived from (1) a day-9 embryo, (2) a day-26 fetus, and (3) mammary gland of a six-year-old ewe in the last trimester of pregnancy. Morphology of the embryo-derived cells is unlike both mouse embryonic stem (ES) cells and the embryo-derived cells used in our previous study. Nuclear transfer was carried out according to one of our established protocols[1] and reconstructed embryos transferred into recipient ewes. Ultrasound scanning detected twenty-one single fetuses on day 50-60 after oestrus (Table 1). On subsequent scanning at ~14-day intervals, fewer fetuses were observed, suggesting either misdiagnosis or fetal loss. In total, 62 percent of fetuses were lost, a significantly greater proportion than the estimate of 6 percent after natural mating.[4] Increased prenatal loss has been reported after embryo manipulation or culture of unreconstructed embryos.[5] At about day 110 of pregnancy, four fetuses were dead, all from embryo-derived cells, and postmortem analysis was possible after killing the ewes. Two fetuses had abnormal liver development, but no other abnormalities were detected and there was no evidence of infection.

Eight ewes gave birth to live lambs (Table 1). All three cell populations were represented. One weak lamb, derived from the few fibroblasts, weighed 3.1 kg and died within a few minutes of birth, although postmortem analysis failed to find any abnormality or infection. At 12.5 percent, perinatal loss was not dissimilar to that occurring in a large study of commercial sheep, when 8 percent of lambs died within 24 h of birth.[6] In all cases the lambs displayed the morphological characteristics of the breed used to derive the nucleus donors and not that of the oocyte donor (Table 2). This alone indicates that the lambs could not have been born after inadvertent mating of either the oocyte donor or recipient ewes. In addition, DNA microsatellite analysis of the cell populations and the lambs at four polymorphic loci confirmed that each lamb was derived from the cell popu-

Table 1. Development of embryos reconstructed with three different cell types

Cell type	No. of fused couplets (%)*	No. recovered from oviduct (%)	No. cultured	No. of morula/ blastocyst (%)	No. of morula or blastocysts transferred†	No. of pregnancies/ no. of recipients (%)	No. of live lambs (%)‡
Mammary epithelium	277 (63.8)[a]	247(89.2)	—	29 (11.7)[a]	29	1/13(7.7)	1 (3.4%)
Fetal fibroblast	172 (84.7)[b]	124(86.7)	—	34 (27.4)[b]	34	4/10 (40.0)	2(5.9%)
			24	13 (54.2)[b]	6	1/6 (16.6)	1 (16.6%)§
Embryo- derived	385 (82.8)[b]	231 (85.3)	—	90 (39.0)[b]	72	14/27 (51.8)	4 (5.6%)
			92	36 (39.0)[b]	15	1/5 (20.0)	0

*As assessed 1 h after fusion by examination on a dissecting microscope. Superscripts a or b within a column indicate a significant difference between donor cell types in the efficiency of fusion ($P < 0.001$) or the proportion of embryos that developed to morula or blastocyst ($P < 0.001$).

†It was not practicable to transfer all morulae/blastocysts.

‡As a proportion of morulae or blastocysts transferred. Not all recipients were perfectly synchronized.

§This lamb died within a few minutes of birth.

lation used as nuclear donor. Duration of gestation is determined by fetal genotype,[7] and in all cases gestation was longer than the breed mean (Table 2). By contrast, birth weight is influenced by both maternal and fetal genotype.[8] The birth weight of all lambs was within the range for single lambs born to Blackface ewes on our farm (up to 6.6 kg) and in most cases was within the range for the breed of the nuclear donor. There are no strict control observations for birth weight after embryo transfer between breeds, but the range in weight of lambs born to their own breed on our farms is 1.2–5.0 kg, 2–4.9 kg, and 3–9kg for the Finn Dorset, Welsh Mountain, and Poll Dorset genotypes, respectively. The attainment of sexual maturity in the lambs is being monitored.

Development of embryos produced by nuclear transfer depends upon the maintenance of normal ploidy and creating the conditions for developmental regulation of gene expression. These responses are both influenced by the cell-cycle stage of donor and recipient cells and the interaction between them (reviewed in ref. 9). A comparison of development of mouse and cattle embryos produced by nuclear transfer to oocytes[10,11] or enucleated zygotes[12,13] suggests that a greater

CLONING

Table 2. Delivery of lambs developing from embryos derived by nuclear transfer from three different donor cells types, showing gestation length and birth weight

Cell type	Breed of lamb	Lamb identity	Duration of pregnancy (days)*	Birth weight (kg)
Mammary epithelium	Finn Dorset	6LL3	148	6.6
Fetal fibroblast	Black Welsh	6LL7	152	5.6
	Black Welsh	6LL8	149	2.8
	Black Welsh	6LL9†	156	3.1
Embryo-derived	Poll Dorset	6LL1	149	6.5
	Poll Dorset	6LL2‡	152	6.2
	Poll Dorset	6LL5	148	4.2
	Poll Dorset	6LL6‡	152	5.3

*Breed averages are 143,147 and 145 days respectively for the three genotypes Finn Dorset, Black Welsh Mountain, and Poll Dorset.

†This lamb died within a few minutes of birth.

‡These lambs were delivered by caesarian section. Overall the nature of the assistance provided by the veterinary surgeon was similar to that expected in a commercial flock.

proportion develop if the recipient is an oocyte. This may be because factors that bring about reprogramming of gene expression in a transferred nucleus are required for early development and are taken up by the pronuclei during development of the zygote.

If the recipient cytoplasm is prepared by enucleation of an oocyte at metaphase II it is only possible to avoid chromosomal damage and maintain normal ploidy by transfer of diploid nuclei,[14,15] but further experiments are required to define the optimum cell-cycle stage. Our studies with cultured cells suggest that there is an advantage if cells are quiescent (ref. 1 and this work). In earlier studies, donor cells were embryonic blastomeres that had not been induced into quiescence. Comparisons of the phases of the growth cycle showed that development was greater if donor cells were in mitosis[16] or in the G1 (ref. 10) phase of the cycle, rather than in S or G2 phases. Increased development using donor cells in G0, G1, or mitosis may reflect greater access for reprogramming factors present in the oocyte cycoplasm, but a direct comparison of these phases in the same cell population is required for a clearer understanding of the underlying mechanisms.

Together these results indicate that nuclei from a wide range of cell types should prove to be totipotent after enhancing opportunities for reprogramming by using appropriate combinations of these cell-cycle stages. In turn, the dissemination of the genetic improvement obtained within elite selection herds will be enhanced by limited replication of animals with proven performance by nuclear transfer from cells derived from adult animals. In addition, gene targeting in livestock should now be feasible by nuclear transfer from modified cell populations and will offer new opportunities in biotechnology. The techniques described also offer an opportunity to study the possible persistence and impact of epigenetic changes, such as imprinting and telomere shortening, which are known to occur in somatic cells during development and senescence, respectively.

The lamb born after nuclear transfer from a mammary gland cell is, to our knowledge, the first mammal to develop from a cell derived from an adult tissue. The phenotype of the donor cell is unknown. The primary culture contains mainly mammary epithelial (over 90 percent) as well as other differentiated cell types, including myoepithelial cells and fibroblasts. We cannot exclude the possibility that there is a small proportion of relatively undifferentiated stem cells able to support regeneration of the mammary gland during pregnancy. Birth of the lamb shows that during the development of that mammary cell there was no irreversible modification of genetic information required for development to term. This is consistent with the generally accepted view that mammalian differentiation is almost all achieved by systematic, sequential changes in gene expression brought about by interactions between the nucleus and the changing cytoplasmic environment.[17]

METHODS

Embryo-derived cells were obtained from embryonic disc of a day-9 embryo from a Poll Dorset ewe cultured as described,[1] with the following modifications. Stem-cell medium was supplemented with bovine DIA/LIF. After eight days, the explanted disc was disaggregated by enzymatic digestion and cells replated onto fresh feeders. After a further seven days, a single colony of large flattened cells was isolated and grown further in the absence of feeder cells. At passage 8, the modal chromosome number was 54. These cells were used as nuclear donors at passages 7–9. Fetal-derived cells were obtained from an eviscerated Black Welsh Mountain fetus recovered at autopsy on day 26 of pregnancy. The head was removed before tissues were cut into small pieces and the cells dispersed by exposure to trypsin. Culture was in BHK 21 (Glasgow MEM; Gibco Life Sciences) supplemented with L-glutamine (2 mM), sodium pyruvate (1mM) and 10 percent fetal calf serum. At 90 percent confluency, the cells were passaged with a 1:2 division. At passage 4, these fibroblast-like cells had modal chromosome number of 54. Fetal cells were used as nuclear donors at passages 4–6. Cells from mammary gland were obtained from a six-year-old Finn Dorset ewe in the last trimester of pregnancy.[18] At passages 3 and 6, the modal chromosome number was 54 and these cells were used as nuclear donors at passage numbers 3–6.

Nuclear transfer was done according to a previous protocol.[1] Oocytes were recovered from Scottish Blackface ewes between 28 and 33 h after injection of gonadotropin-releasing hormone (GnRH), and enucleated as soon as possible. They were recovered in calcium- and magnesium-free PBS containing 1percent FCS and transferred to calcium-free M2 medium[19] containing 10 percent FCS at 37°C. Quiescent, diploid donor cells were produced by reducing the concentration of serum in the medium from 10 to 0.5 percent for five days, causing the cells to exit the growth cycle and arrest in G0. Confirmation that cells had left the cycle was obtained by staining with antiPCNA/cyclin antibody (Immuno Concepts), revealed by a second antibody conjugated with rhodamine (Dakopatts).

Fusion of the donor cell to the enucleated oocyte and activation of the oocyte were induced by the same electrical pulses, between 34 and 36 h after GnRH injection to donor ewes. The majority of reconstructed embryos were cultured in ligated oviducts of sheep as before, but some embryos produced by transfer from embryo-derived cells or fetal fibroblasts were cultured in a chemically defined medium.[20] Most embryos that developed to morula or blastocyst after six days of culture were transferred to recipients and allowed to develop to term (Table 1). One, two, or three embryos were transferred to each ewe depending upon the availability of embryos. The effect of cell type upon fusion and development to morula or blastocyst was analyzed using the marginal model of Breslow and Clayton.[21] No comparison was possible of development to term as it was not practicable to transfer all embryos developing to a suitable stage for transfer. When too many embryos were available, those having better morphology were selected.

Ultrasound scan was used for pregnancy diagnosis at around day 60 after oestrus and to monitor fetal development thereafter at two-week intervals. Pregnant recipient ewes were monitored for nutritional status, body condition, and signs of EAE, Q fever, border disease, louping ill, and toxoplasmosis. As lambing approached, they were under constant observation and a veterinary surgeon called at the onset of parturition. Microsatellite analysis was carried out on DNA from the lambs and recipient ewes using four polymorphic ovine markers.[22]

NOTES

1. K. H. S. Campbell, J. McWhir, W. A. Ritchie, and I. Wilmut, "Sheep Cloned by Nuclear Transfer from a Cultured Cell Line," *Nature* 380 (1996): 64–66.

2. D. Solter, "Lambing by Nuclear Transfer," *Nature* 380 (1996): 24–25.

3. J. B. Gurdon, R. A. Laskey, and O. R. Reeves, "The Developmental Capacity of Nuclei Transplanted from Keratinized Skin Cells of Adult Frogs," *Journal of Embryology and Experimental Morphology* 34 (1975): 93–112.

4. T. D. Quinlivan, C. A. Martin, W. B. Taylor, and I. M. Cairney, "Pre- and Perinatal Mortality in Those Ewes that Conceived to One Service," *Journal of Reproduction and Fertility* 11 (1966): 379–90.

5. S. K. Walter, T. M. Heard, and R. F. Seamark, "In Vitro Culture of Sheep Embryos without Co-culture: Successes and Perspectives," *Theriogenology* 37 (1992): 111–26.

6. M. L. Nash, L. L. Hungerford, T. G. Nash, and G. M. Zinn, "Risk Factors for Perinatal and Postnatal Mortality in Lambs," *Veterinary Record* 139 (1996): 64–67.

7. G. E. Bradford, R. Hart, J. F. Quirke, and R. B. Land, "Genetic Control of the Duration of Gestation in Sheep," *Journal of Reproduction and Fertility* 30 (1972): 459–63.

8. A. Walton and J. Hammond, "The Maternal Effects on Growth and Conformation in Shire Horse-Shetland Pony Crosses," *Proceedings of the Royal Society*, Series B, B125 (1938): 311–35.

9. K. H. S. Campbell, P. Loi, P. J. Otaegui, and I. Wilmut, "Cell Cycle Co-ordination in Embryo Cloning by Nuclear Transfer," *Reviews in Reproduction* 1 (1996): 40–46.

10. H.-T. Cheong, Y. Takahashi, and H. Kanagawa, "Birth of Mice after Transplantation of Early-cell-cycle-stage Embryonic Nuclei into Enucleated Oocytes," *Biology of Reproduction* 48 (1993): 958–63.

11. R. S. Prather et al., "Nuclear Transplantation in the Bovine Embryo: Assessment of Donor Nuclei and Recipient Oocyte," *Biology of Reproduction* 37 (1987): 859–66.

12. J. McGrath and D. Solter, "Inability of Mouse Blastomere Nuclei Transferred to Enucleated Zygotes to Support Development In Vitro," *Science* 226 (1984): 1317–18.

13. J. M. Robl, et al., "Nuclear Transplantation in Bovine Embryos," *Journal of Animal Science* 64 (1987): 642–47.

14. K. H. S. Campbell, W. A. Ritchie, and I. Wilmut, "Nuclear-cytoplasmic Interactions during the First Cell Cycle of Nuclear Transfer Reconstructed Bovine Embryos: Implications for Deoxyribonucleic Acid and Development," *Biology of Reproduction* 49 (1993): 933–42.

15. F. L. Barnes et al., "Influence of Recipient Oocyte Cell Cycle Stage on DNA Synthesis, Nuclear Envelope Breakdown, Chromosome Constitution, and Development in Nuclear Transplant Bovine Embryos," *Molecular Reproduction and Development* 36 (1993): 33–41.

16. O. Y. Kwon and T. Kona, "Production of Identical Sextuplet Mice by Transferring Metaphase Nuclei from 4-cell Embryos," *Journal of Reproduction and Fertility*, Abstract Service 17 (1996): 30.

17. J. B. Gordon, *The Control of Gene Expression in Animal Development* (Oxford: Oxford University Press, 1974).

18. L. M. B. Finch et al., "Primary Culture of Ovine Mammary Epithelial Cells," *Biochemical Society Transactions* 24 (1996): 369S.

19. W. K. Whitten and J. D. Biggers, "Complete Development In Vitro of the Preimplantation Stages of the Mouse in a Simple Chemically Defined Medium," *Journal of Reproduction and Fertility* 17 (1968): 399–401.

20. D. K. Gardner, M. Lane, A. Spitzer, and P. A. Batt, "Enhanced Rates of Cleavage and Development for Sheep Zygotes Cultured to the Blastocyst Stage in In Vitro in the Absence of Serum and Somatic Cells: Amino Acids, Vitamins, and Culturing Embryos in Groups Stimulate Development," *Biology of Reproduction* 50 (1994): 390–400.

21. N. E. Breslow and D. G. Clayton, "Approximate Inference in Generalized Linear Mixed Models," *Journal of the American Statistical Association* 88 (1993): 9–25.

22. F. C. Buchanan, R. P. Littlejohn, S. M. Galloway, and A. L. Crawford, "Microsatellites and Associated Repetitive Elements in the Sheep Genome," *Mammalian Genome* 4 (1993): 258–64.

ACKNOWLEDGMENTS

We thank A. Colman for his involvement throughout this experiment and for guidance during the preparation of this manuscript; C. Wilde for mammary-derived cells; M. Ritchie, J. Bracken, M. Malcolm-Smith, W. A. Ritchie, P. Ferrier, and K. Mycock for technical assistance; D. Waddington for statistical analysis; and H. Bowran and his colleagues for care of the animals. This research was supported in part by the Ministry of Agriculture, Fisheries and Food. The experiments were conducted under the Animals (Scientific Procedures) Act 1986 and with the approval of the Roslin Institute Animal Welfare and Experiments Committee.

3) The Birth of Cloning

J. B. Gurdon

The first sentence in the now-famous *Nature* article announcing the birth of a cloned sheep mentions an experiment in which "keratinocytes" from adult amphibians developed into tadpoles. The reader studious enough to consult the footnotes would have learned that Ian Wilmut and his colleagues at the Roslin Institute and PPL Therapeutics in Scotland were referring to experiments that two colleagues and I did at the Laboratory of Molecular Biology in Cambridge, England, twenty-two years ago. We injected the nuclei of skin cells (keratinocytes) from the foot webs of adult frogs into frog eggs from which the nuclei had been removed. Some of those eggs developed into swimming tadpoles.

The salient points of that history are all implicit in the first sentence of the *Nature* article. The first is that the experiments of Wilmut and his colleagues *have* a history; the cloning of an adult sheep was the latest in a series of experiments that dates back to the 1950s. The second is that the cloning of a sheep required a

This article is reprinted by permission of *The Sciences* and is from the September/October 1997 issue, pages 26–31. The Sciences, 2 East 63rd Street, New York, NY 10021.

CLONING

great improvement over previously available techniques. Whereas my colleagues and I had generated frogs at various juvenile stages of development from the nuclei of adult cells, Wilmut and his colleagues induced the nucleus of an adult cell to develop into what, presumably, will grow into an adult vertebrate. The third historical point is simply that the extension of nuclear-transplantation from amphibians to mammals did not happen overnight. The first successful nuclear-transfer experiments with frog cells were done in 1952; the first confirmation of successful nucleartransfer experiments with mice came in 1983.

How does a seemingly structureless egg convert itself in short order into a complex and highly organized animal? The question has riveted scientists and natural philosophers since the time of Aristotle. But only recently have biologists developed the technical means to answer it.

One imaginative solution, popularized by the Swiss naturalist Charles Bonnet in the 1700s, was the theory of preformation. Bonnet suggested that each egg contains a miniature embryo, which in turn contains an ovary with eggs, each of which contains its own miniature embryo complete with ovaries and eggs, and so on. In other words, each new generation nests inside the preceding one, much like a set of Russian babushka dolls. But even Bonnet did not fully believe his theory because it implied, when taken to its logical conclusion, that Eve must have had 27 million embryos in her ovaries. And preformation was eventually discredited by the eighteenth-century German embryologist Kaspar Friedrich Wolff, who observed that in both plants and animals specialized organs arise from unspecialized tissue.

Wolff's observations suggested instead that the egg held a mysterious "essential force" capable of orchestrating development. By the 1870s and 1880s, the German biologist August Weismann had launched the idea that the real essence of life is the germ plasm in eggs and sperm. That plasm, he asserted, periodically grows an organism around itself as a device to produce more eggs and sperm. As Samuel Butler, the Victorian author of *The Way of All Flesh*, put it: "A hen is only an egg's way of making another egg." Weismann also suggested that the germ plasm lived in what were subsequently called chromosomes—literally, "colored bodies."

But that theory, called epigenesis, still left a lot to be explained. In particular, how could the chromosomes in one cell give rise to many dissimilar cells? By what mechanism did some cells become muscle cells whereas others became gut or skin cells? In 1892 Weismann suggested the existence of different kinds of nuclei, sequestered in different parts of the embryo during cell division, where they then guided the development of different kinds of cells.

Weismann's theory, as it turned out, was not consistent with experimental evidence. Early in this century the German zoologist Hans Spemann ran a series of experiments in which he tied a thin hair—he recommended the delicate hair from a blond infant less than nine months old—around a fertilized newt egg, creating a lobe of cytoplasm. At the eight-cell stage of division, he loosened the knot enough to allow one of the nuclei from the developing embryo to move into the lobe's cytoplasm. He discovered that nuclei at the eight-cell stage were capable of directing the development of a normal embryo, a finding that implied that all of the newt embryo's nuclei, at least up to the eight-cell stage, were genetically equivalent. Spemann also found, however, that if the same experiment was carried out later in development, the tied-off portion would generate only part of an embryo. It appeared that between the early and later stages of growth, cells lost their "potency" and as a result, their developmental potential became fixed.

As Spemann was aware, cellular fates might be sealed in at least two ways. The identity of individual cells might be established by their local environment—their positions relative to other tissues. But it was also possible that during development the nucleus underwent changes that progressively stifled its developmental capacity, making it less than totally potent.

Spemann suggested that the developmental status of the nucleus could be checked by transferring it to an egg from which the nucleus had been removed. Presumably, if the transferred nucleus was still unmodified, it would give rise to a normal embryo. If it was modified, the egg would die or develop abnormally. For the next fifty years biologists struggled to discover via nuclear-transplant experiments whether, in the course of development, nuclei remained totipotent (capable of producing a fertile adult), pluripotent (capable of giving rise to various kinds of embryonic tissue but not to a fertile adult), or nullipotent (unable to direct the development of cells unlike themselves).

The first real success in nuclear transplantation in a vertebrate was reported in 1952 by Robert Briggs and Thomas J. King. Working at the Institute for Cancer Research and Lankenau Hospital Research Institute in Philadelphia, Pennsylvania, the two embryologists developed a procedure for transplanting a nucleus from an embryo of the leopard frog, *Rama pipiens*, into a *Rana* egg. For this early work, whose purpose was to demonstrate the technique rather than to test the developmental status of nuclei, Briggs and King transplanted nuclei from cells they believed to be undifferentiated. (The procedure itself is described in "Backward Compatible," by Marie A. Di Berardino and Robert G. McKinnell, page 32.)

To better understand this and subsequent experiments in nuclear transplantation, one needs to be familiar with the basic steps in the growth of an animal embryo. The first step in development from a single fertilized egg cell is a series of consecutive cell divisions, or cleavages. The cleavages give rise to a hollow ball of cells called a blastula. The cells then rearrange themselves into a three-layered structure called a gastrula. The outer layer of the gastrula is the ectoderm, the middle layer is the mesoderm and the inner layer is the endoderm. After the three germinal layers form, the neural tube, from which the entire nervous system and the head will develop, begins to take shape. At this stage the embryo is called a neurula. After neurulation, a complicated series of cell interactions gives rise to the body's organ systems.

The nuclei from undifferentiated cells with which Briggs and King first experimented came from late-stage blastulas. Out of 197 nuclei the workers transferred, only a few cleaved properly and then developed into complete blastulas—though most of the blastulas did grow into tadpoles. Under the difficult technical circumstances, Briggs and King considered their limited outcome a success. "This means that the nuclei are not significantly damaged," they wrote, "and indicates that the technique of nuclear transplantation may now be used in testing nuclei from various differentiated parts of the Amphibian embryo."

They took their own advice. Briggs and King tried transplanting nuclei from successively more advanced embryos into eggs. In 1953 they successfully transplanted nuclei from early-stage gastrula cells. In 1954 they reported success, albeit more limited, with late-stage gastrula nuclei. Only 8 or 9 percent of the eggs in the 1954 experiment cleaved normally; of the ones that did, most were arrested in various neurula and post-neurula stages, never squirming into life as tadpoles. But as Briggs and King later wrote, "at the time we could not eliminate the possibility that the developmental deficiencies might be the result of nuclear damage during the operation. Therefore, emphasis was placed on the positive result...."

Beginning to suspect that their technique was not their only hurdle, Briggs and King transplanted nuclei removed from embryos at a wide range of stages of growth, including stages more advanced than any they had worked with before. They chose endoderm cells for the test because, of the three germinal layers, the endoderm is determined earliest in development and is made up of large cells that are easy to handle.

Those experiments led to two troubling results. First, the transplants gave rise to many abnormal embryos—and the abnormalities seemed consistent. The ectoderms of the embryos were poorly differentiated, whereas the endoderms and mesoderms were nearly normal. Since the donor nuclei originated from endoderm cells, the abnormalities seemed to be coming from normal cell differentiation and not from damage during transplantation.

The second result was more mundane but equally troubling. As nuclei were taken from progressively differentiated endoderm cells, a progressively smaller proportion of the nuclei led to normal cleavage in the eggs to which they had been transplanted. By the same token, a progressively smaller proportion of the transplanted nuclei could promote the development of normal tadpoles. Among completely cleaved blastulas of normal appearance, 20 percent of the ones derived from late-gastrula endoderm nuclei developed into tadpoles, but only one blastula among the ones that were derived from mid-neurula or from tail-bud donors grew into a tadpole. For all those reasons, Briggs and King concluded that the nuclear-transplant experiments with endoderm tissue showed "that endoderm nuclei undergo stabilized or 'irreversible' changes during differentiation."

It was at that point that I entered the picture. In 1956 I began graduate work at the University of Oxford under the supervision of the developmental biologist Michael Fischberg. There I embarked on nuclear-transplant experiments in *Xenopus laevis,* the South African clawed frog.

Thanks to a discovery of Fischberg's, those experiments were particularly well controlled. In the early transplant work, it was a tricky business to make sure the nucleus of the egg had been replaced by that of the donor cell. Fischberg had discovered a *Xenopus* mutant that had only one nucleolus in its cells instead of the normal two. (A nucleolus is a small nuclear body made up mostly of protein.) The discovery was important because the single nucleolus could serve as a marker, making it possible to determine beyond a doubt that in nuclear-transplant experiments the egg nucleus had been removed. When a donor nucleus with one nucleolus was transplanted into an enucleated egg of the normal two-nucleoli strain, the emergence of an embryo whose cells had one nucleolus proved that the original egg nucleus had been killed.

Our early experiments led to two major results. First, as I reported in 1958 with Fischberg and Thomas R. Elsdale, we were able to rear nuclear-transplant embryos to sexually mature adults, not just to the tadpole stage, as had been done with *Rana.* That was the first demonstration that the nuclei of cells that had undergone at least some development remained totipotent, not just pluripotent. Second, we were able to get normal development with much older donor tissue than had been reported for *Rana.* In 1960 I reported that normal nuclear-transplant embryos, including some that developed into feeding tadpoles, had been grown from endoderm cells of tail-bud tadpoles, not just from the cells of gastrulas. Thus our work with *Xenopus* led us to conclude that the specialization of cells is not necessarily accompanied by a loss or deadening of gene expression—a conclusion that differed from the one reached by Briggs and King.

How could those different results be explained? If changes in the nucleus play a role in triggering cell differentiation, one would expect those changes to arise consistently at certain stages of development, no matter what the species. So why were the experimental results with *Xenopus* different from those with *Rana?* And which set of experiments was the better guide to the problem of cell differentiation?

We had confidence in our results with *Xenopus,* for the following reason. Generally, positive outcomes are more significant than negative ones. It was hard to imagine how *Xenopus* nuclei could be permanently altered during development and then suddenly change to promote normal growth. But it was not hard at all to imagine how potent *Rana* nuclei might be changed by the manipulations the experiments required. For that reason, one swimming tadpole carried greater weight with us than many arrested embryos.

In time, an alternate explanation for Briggs and King's puzzling results began to emerge. Although early articles described chromosomal defects in only the most abnormal embryos, more detailed studies subsequently done by the developmental biologist Marie A. Di Berardino of the Medical College of Pennsylvania-Hahnemann School of Medicine, Allegheny University of the Health Sciences in Philadelphia clarified the situation. Di Berardino showed that all the embryos that died before the feeding-tadpole stage of development had visibly abnormal chromosomes. Moreover, the more severe the developmental disorder, the more severe the prior chromosomal defects. Thus the reason workers had been able to obtain pluripotent embryonic nuclei at a more advanced stage of *Xenopus* than of *Rana* seemed to be that the *Xenopus* nuclei or chromosomes are more resistant to damage during transplantation.

The nature of the chromosomal abnormalities suggested that at least some of them were caused by a lack of synchrony between the cell-division cycle of the egg and that of the donor cell. Before the egg cleaves, the nucleus must create a duplicate set of chromosomes, one set for each daughter cell. Because the egg cytoplasm largely determines when a transplanted nucleus first replicates its chromosomes and divides, donor nuclei could be forced to divide so soon after transplantation that they might not have had time to replicate all their chromosomes. Premature cleavage would then give rise to daughter cells with chromosomal defects.

Eventually, investigators found ways to partially "derepress" the donor nucleus; chromosomal replication was then more likely to be completed within the first cycle of the egg. In 1970 the developmental biologist Sally Hennen of Marquette University in Milwaukee, Wisconsin, reported that adding a protein called spermine to the nuclear-transfer medium and performing the transplant at a lower temperature greatly improved the results of nuclear-transplant experiments with nuclei from the endoderm of *Rana*. In 1983 Di Berardino and the developmental biologist Nancy Hoffner Orr of the Allegheny University of the Health Sciences achieved much the same improvements by transplanting donor nuclei into oocytes, immature egg cells from the ovary, rather than into mature egg cells that had already been released from the ovary. Nuclei transplanted into an oocyte had twenty-four hours' exposure to the cytoplasm of the oocyte before cleavage, much longer than did nuclei transplanted into mature eggs; therefore, in oocytes there was enough time for the cytoplasm to reset the cell-division cycles. Those techniques led to transplantation successes with *Rana* comparable to those obtained with *Xenopus.*

In the meantime, those of us working with *Xenopus* had been attempting transplants with nuclei from more advanced embryos. In the absence of the development of a normal adult frog, the success of a nuclear-transplant experiment is judged in part by the developmental stage the embryo reaches. By 1962 I had succeeded in growing normal feeding tadpoles from the nuclei of cells lining the intestine of feeding tadpoles. The intestinal cells had striated borders, which revealed their differentiated state; that was the first clear evidence that the nuclei of differentiated cells are pluripotent.

By 1966 I had obtained adult frogs from cells taken from tadpole intestines. In all, 120 transplanted intestinal nuclei yielded seven adult frogs, five of which were fertile. That experiment was perhaps the single most important one we performed, because it proved that a cell can undergo specialization and yet remain totipotent, retaining all the genetic material needed to make a complete, sexually mature individual. This key experiment justified the view that the cloning of differentiated, and perhaps even adult, cells was at least theoretically possible.

If terminally differentiated cells from a juvenile stage could give rise to fertile adults, albeit only occasionally, why not cells from the various tissues of the adults themselves? In later experiments the molecular biologist Ronald A. Laskey of the Laboratory of Molecular Biology in Cambridge and I grew the cells of adult animals in culture and then took nuclei from those cells for transplantation. Whether the cells came from the adult's kidney, heart, skin, or lung, they generated fully formed tadpoles. The cell cultures, however, seemed to be made up largely of fibroblasts, cells that are not obviously specialized but that tend to proliferate when tissues are cultured. For that reason we could not rule out the possibility that the tadpoles had come from cell nuclei that were not terminally differentiated.

In 1975 we had further success. That year Laskey, Raymond Reeves and I wrote the paper to which Wilmut would refer twenty-two years later. In the *Journal of Embryology and Experimental Morphology* we reported the transplantation of nuclei from cells derived from adult frogs. We used skin cells from the adults' foot webs. When small pieces of the foot web are cultured, a single layer of cells grows outward from the excised tissue. We treated those monolayers with an antibody that binds to keratin, a protein made only by specialized cells, and we found that 99.9 percent of the cells in the monolayers bound to the antibody, thus revealing that they contained keratin. Nuclear-transplant experiments done with those cells yielded heartbeat-stage tadpoles with well-differentiated eyes and other organs. The probability that those clones originated from the 0.1 percent of

CLONING

the donor-cell population unproved to contain keratin was less than one in 10 billion. We felt justified in concluding that "cell specialization does not involve any loss, irreversible inactivation, or permanent change in chromosomal genes."

At this point you may be wondering how our experiments fell short of Wilmut cloning of a cell from the udder of an adult sheep. From one perspective the answer is simple. In all our experiments we were never able to generate an adult animal from the nuclei of cells derived from an adult animal. We had produced fertile adult frogs from the nuclei of tadpole intestinal cells, but those cells came from a juvenile rather than an adult animal. And we had grown heartbeat-stage tadpoles from the nuclei of adult skin cells, but none of our skin-cell embryos survived to become adult frogs. Wilmut was the first to generate an adult (or soon-to-be adult) animal from the nucleus of an adult cell.

From the perspective of developmental biology, however, the distinction between our experiments—particularly the one with the tadpole intestinal cells—and Wilmut's is fairly subtle. In the 1966 intestinal-cell experiment we grew frogs from cells we knew to be terminally differentiated, even if they came from tadpoles. Wilmut and his colleagues, on the other hand, grew Dolly from what they knew to be an adult cell, but they could not know for sure that it was terminally differentiated. In their *Nature* article they remark that they cannot exclude the possibility that Dolly developed from a stem cell, a relatively undifferentiated kind of cell capable of giving rise to other kinds of cells. But the experiments complement one another, both of them demonstrating that development does not mean loss of nuclear potency.

Wilmut's experiment is intriguing primarily because he was able to clone an adult animal from the cell of another adult, but it is also significant because of what it says about the problem of aging. Biologists have long thought that cell lineages are mortal, or in other words that a cell can divide only a limited number of times before it fads and dies. In early experiments with *Xenopus*, cells from a late-blastula-stage embryo that grew from an egg with a transplanted nucleus were themselves sometimes transplanted, in serial experiments. Each transplanted nucleus underwent twelve divisions between the egg and the late-blastula stage, and serial-transplant experiments sometimes continued for six or seven generations. Thus in some experiments the nucleus underwent seventy to eighty divisions, more than frog cells undergo in the normal lifespan of a frog. Wilmut's experiment seems to extend this result, showing that there is no intrinsic aging process that necessarily eliminates the totipotency of nuclei (see "Born Again?" by Ronald Hart, Angelo Turturro, and Julian Leakey, page 47).

Why did the skin-cell embryos in our experiments die before maturity? We do not know, but the likeliest explanation is the forced pace of the initial cell divisions. The cells of an adult animal divide infrequently, if ever. When the nucleus of an adult cell is injected into a frog's egg, however, it is forced to undergo its first division on the egg's schedule. And eggs normally divide about eighty minutes after being injected with a nucleus. So it is hardly surprising that replication cannot always be finished, and that the older the embryo or the animal from which the donor nucleus is taken, the more likely it is that the chromosomes and genes will be incompletely replicated and, consequently, that the growth of the embryo will be abnormal.

Indeed, timing problems may account for the varying difficulty with which frogs, mice, and sheep can be cloned. Mammalian embryos have slower cell-division rates than amphibian embryos have, and that is probably one reason some mammals are easier to clone than amphibians: the egg cytoplasm has more time to reprogram the nuclear DNA. Mice have proved difficult to clone—only cells from very early mouse embryos have yielded nuclei capable of promoting normal development—but workers have had greater success with sheep. That is probably because the protein products of nuclear genes must appear by the two-cell stage in the development of the mouse embryo, whereas in sheep they are needed only at the eight-cell stage or later.

In general, however, I believe that some genetic damage is hard to avoid when nuclei from the cells of advanced embryos or adults are transplanted. That damage probably accounts for the failure until now to create a normal adult animal from the cells of another adult animal. Still, that does not change how my colleagues and I interpret our experiments. They show, we believe, that cell differentiation does not involve any permanent or stable change in the chromosomes. In general, differentiation takes place through the controlled expression of genes, rather than via their loss or irreversible inactivation. That developmental fact is the foundation on which Dolly stands, and it makes the cloning of other adult vertebrates at least theoretically possible.

BIOENGINEERING:
ANIMALS AND PLANTS

Before we get to the question of human cloning, there are issues concerning animal and plant cloning. We begin this section with three pieces extolling the wonders and virtues of cloning. First, we have a tale full of inspiration and hope about that mammal with the world's biggest sex appeal: the panda. Warm and cuddly (at least in popular imagination), the panda is the symbol of threatened species. This is not just because of human interference—poaching and destruction of habitat—but because of its own peculiarities. It has a distinctive and demanding diet and, even worse, little ability or apparently the inclination to reproduce. Many a zookeeper has spent hours trying to get pandas to do what, for other species, comes naturally. Now, however, with cloning, new possibilities open up. Perhaps we can get around both nature and the animals themselves and produce pandas by the hundreds—at least this is the message from Claire Cockcroft.

Next, somewhat more soberly but with no less enthusiasm, Harry M. Meade tells us about his work in using animals to produce valuable pharmaceutical products. He regards cloning as a terrific boon in producing animals of the kind that he needs: made to order and in quick time. There is no need to wait on the delays and vagaries of natural breeding. To him, an animal is a tool or machine like any other, and cloning is a way of improving its efficiency. Following this comes "Brave New Rose"—note the play on the title of Aldous Huxley's book—which deals with the possibilities and opportunities opened by cloning and related practices in the plant world. The forestry industry, for instance, will probably be transformed by cloning, as trees with and only with precisely the desired properties are grown. Fast-growing pulp trees for the newspaper business can be produced to order. The English poet Rupert Brooke once wrote a poem bemoaning the fact that in Germany "even tulips grow as they are told." We doubt he would find much joy in northern Canada in the next century.

From pandas to poplars. Already a somewhat darker note is starting to be struck. Therefore, it is appropriate to conclude this section with an article dealing with some of the ethical and related questions raised by cloning techniques in the nonhuman living world. B. Rollin argues that there are three basic socioethical issues that must be tackled. First, those associated with a fear of the unknown. These he does not think are serious. Second, there are those associated with risks, to us and the environment and to other organisms. He does not think that these should be minimized, especially those concerned with a narrowing of the basic gene pool or reservoir on which one can draw. Third, there are issues about the potential pain or suffering of genetically engineered animals. Here Rollin feels that there may be significant questions although not necessarily unanswerable questions. Somewhat pessimistically he states: "I doubt that cloning will worsen the situation we already have—though it won't improve it either."

It is true that Rollin is concerned primarily with issues in the animal world, but you can see easily that some of these apply directly to the plant world, as do his comments and suggestions. The issue of pain does not arise, but the question of depletion of variety is if anything even more pressing. We are all concerned about the possible loss of a species like the panda. Few of us give any thought to, say, the loss of a species or subspecies of bush or shrub or similar plant. Yet the loss, perhaps through neglect brought on by enthusiasm for the cloning of instantly desirable plants, could come back to haunt us. Also, as will be brought home strongly in the discussion of cloning by religious leaders and thinkers, the trend toward uniformity anywhere in the living world raises serious ethical and theological questions.

4 Pandas and Cloning

Claire Cockcroft

Giant pandas have roamed the misty mountains of China for half a million years but their once extensive habitat is, today, pitifully small. Fewer than one thousand pandas live in the bamboo-rich forest fragments covering the mountains of southwest China, and they could be extinct within twenty-five years.

Breeding programs and artificial insemination are delaying their demise.

Developments in biotechnology, however, mean that cloning is now a realistic alternative for saving these animals teetering on the edge of extinction.

For the panda, with its fussy eating habits and reluctance to reproduce, duplication could be the most pragmatic approach.

Pandas have particularly poor powers of procreation, preferring to eat or sleep off the effects of bamboo overindulgence than to engage in sexual practices. Timing is crucial, as females usually produce eggs once a year and are only fertile for a few days—enough to try the patience of any zoologist trying to breed them.

They live in small isolated groups in the wild and inbreeding has undoubtedly contributed to their high infertility rate.

CLONING

Just 10 percent of males are capable of mating and only 30 percent of females ovulate naturally, according to the China panda breeding-technology committee.

About one hundred pandas are in zoos worldwide. They are notoriously difficult to breed in captivity and cubs often do not survive.

The first success story was in the Beijing zoo in 1963, despite the initial confusion in determining the sex of the animal they were attempting to mate.

Several panda cubs have been born in breeding centers in China this year, including the first set of triplets but the long-term survival rate is low.

Pandas are also threatened by poaching and the loss of their natural habitat, because of the demand for land to house and feed China's growing population. Their habitat has shrunk to half the size it was in 1984 when pandas were listed as an endangered species.

Food is another problem. Pandas have the digestive system of a carnivore but are vegetarians at heart. Their highly specialized diet consists almost exclusively of two species of bamboo—they spend up to fourteen hours a day munching their way through 20 to 40 pounds.

The unusual life cycle of their major food source is another problem. Every thirty to one hundred years most bamboo plants in an area will decide to flower and then die off after dropping their seeds. It takes around a year for new plants to grow from seed but about twenty years' growth to support a panda population.

In the past, pandas would simply amble away to a new region if food became scarce. This is no longer possible as their islands of bamboo forest are separated by farmland or housing. Some pandas have starved to death.

To preserve the planet's endangered species, genes of rare plants and animals are being assembled in "frozen zoos" in laboratories around the world. These twentieth-century "Noah's arks" contain the information from which these species may be cloned in the future. While cloning could restore the numbers of endangered species, some scientists are worried that it would produce a population lacking genetic diversity, which in itself would be a threat to survival.

The Chinese strongly believe in preserving these enigmatic bears once treasured by their emperors. The arrival of Dolly the sheep prompted the authorities to initiate a panda cloning-research program. As fertile female pandas are scarce, and panda eggs are preferentially used for artificial insemination programs, they have pioneered the use of eggs from other species.

Earlier this year, the team, led by Prof. Chen Dayuan, announced the successful cloning of a panda embryo. The genetic information from the cells of a dead female panda were introduced into the eggs of a Japanese white rabbit from which the nucleus had been removed. For the gestation of this embryo, a suitable surrogate species, with similar genetic makeup, has to be found.

Mr. Chen believes his best chances of success lie in finding a non-panda surrogate. His team is looking for a "mother" that is large enough and has reproductive traits similar to that of a giant panda.

Two possible candidates are the black bear and the sloth bear. But previous research on growing embryos of one species in the womb of another suggests Mr. Chen's task is going to be difficult. The project is further complicated by the fact that the fundamental biology of pandas remains poorly understood.

If cloning is to save the giant panda it will have to be quick. But for researchers planning to clone less-endangered animals there is more time to iron out technical problems.

The success rate in obtaining a live birth from a cloned embryo is very low: 276 attempts were required to produce Dolly. And it is not known what the effects of using a surrogate egg will be on the clone.

Nevertheless, scientists hope that the first panda clone will make its debut early in the next millennium. Cloning has already been used to save other endangered species.

In New Zealand, researchers have produced Elsie, the world's second cloned cow—confirming that the procedure works. What makes the calf unique, however, is that it originates from a cow that is the last known member of its breed.

The breed is peculiar to one of the Auckland Islands, a cold, windswept group about 400 kilometers south of New Zealand. The cow, whose name is Lady, is a leftover from attempts to farm the islands last century.

Lady is unique, says Dr. David Wells of New Zealand's Ag Research Station, one of a tough breed that adapted to harsh conditions by developing a taste for seaweed. "We're interested in establishing the breed again to see whether there are any useful traits," he says.

Taking the nucleus from one of Lady's cells, Dr. Wells inserted Lady's DNA into another cow's egg from which the nuclear DNA had been removed. He then implanted the resulting embryo into another cow's womb and on July 31, Elsie was born. To be certain that she was, indeed, a clone of Lady, Dr. Wells sent samples of DNA from Lady, Elsie, and from the cells taken from Lady used to create Elsie to another lab in New Zealand and asked that lab to compare the samples.

"They found no differences," he says. "They were all genetically identical . . . Elsie is a genetic replica of Lady."

Dr. Wells says that to have any hope of success in cloning, you have to know a lot about the embryology and reproductive biology of the animal you want to clone. You have to have a good supply of eggs, and you have to have a good supply of surrogate mothers.

"Clearly, all those requirements, you know, were available to us in cloning

CLONING

Lady. Her being a cow, we knew enough about the reproductive biology. We could use other eggs from other cows for nuclear transfer."

Meanwhile, in Texas, researchers are hard at work on the Missyplicity experiment. The $2.3-million (U.S.) project is an attempt to clone a dog called Missy, now dead. Funding is from the former owners of the animal who wish to clone their old pet and generally understand more about cloning. And in Australia, scientists are contemplating cloning the extinct Tasmanian tiger.

A woolly mammoth, preserved within the layers of Siberian permafrost for 23,000 years, might be resurrected by cloning. The nucleus of a woolly mammoth cell would be injected into an elephant egg and the cloned embryo implanted into a surrogate elephant mother, who would carry and give birth to the mammoth clone.

Although cloning has the potential to restore the numbers of endangered species, it does not solve the problem of habitat destruction. As humans have upset the balance, they have a moral responsibility to preserve species diversity. Combining biotechnological methods with a better understanding of animal reproduction and improved environmental management could help to accomplish this.

Without the intervention of science, endangered species face an uncertain future.

5 Dairy Gene

Harry M. Meade

Years ago, when I was a graduate student in biology at MIT, I never dreamed I would end up running a goat farm. But as a vice president at Genzyme Transgenics Corporation in Framingham, Massachusetts, I help oversee barns, a state-of-the-art animal hospital and clinic, and a herd of more than five hundred dairy goats on 168 rural acres in nearby Charlton. Ours is no ordinary farm, however: some of our goats are transgenic animals; in other words, their genetic structure has been altered by human intervention. Although the animals look and act like others of their species, they are being engineered to produce human proteins in their milk. Those proteins include antithrombin III, which is used to prevent blood clotting, and alpha-1-antitrypsin, which prevents lung scarring in people whose bodies do not make enough of that protein.

It has been more than a decade since a colleague and I had the idea of making dairy animals produce pharmaceuticals. I was working at Biogen, Inc., a biotechnology company in Cambridge, Massachusetts. Late one night, while we were

This article is reprinted by permission of *The Sciences* and is from the September/October 1997 issue, pages 20–25. The Sciences, 2 East 63rd Street, New York, NY 10021.

working in the lab, Nils Lonberg (now at GenPharm International, Inc., in Palo Alto, California) and I were discussing the new technique of making transgenic mice that had human proteins in their blood. Having grown up on a dairy farm, I was struck by a new thought: get animals to produce special proteins in their milk instead. That way we would not have to bleed them; we could just milk them twice a day and then purify the human proteins out of their milk. I was reminded of the fable about the goose that laid a golden egg each day: killing the goose left the owner with nothing, while caring for it patiently would have yielded a lifetime of steady benefits.

Innovative ideas tend to arise spontaneously in more than one quarter. Scientists at another Massachusetts company, Integrated Genetics, Inc., also began working on this method. Integrated Genetics later became part of the Genzyme Corporation in Cambridge, of which Genzyme Transgenics is a spin-off. So now we at Genzyme Transgenics are among the pioneers of a new industry. Two other companies are also in the business: Pharming Holding in the Netherlands and PPL Therapeutics in Edinburgh, Scotland.

So far none of the three companies has put medicines on the market. Genzyme Transgenics is furthest along, with antithrombin III, which has undergone safety trials for the U.S. Food and Drug Administration and now must be tested for efficacy. Our company is in the advanced development stages with alpha-1-antitrypsin (also known as alpha-1 proteinase inhibitor), and we have also been able to produce two monoclonal antibodies, which would deliver medicines directly to tumors. Those antibodies are now being evaluated by the clients who commissioned us to make them. Bristol-Myers Squibb Company in New York City, and a Japanese pharmaceutical company that has requested anonymity.

Many protein-based drugs, including insulin, for diabetes; interferon, an antiviral agent; Factors VIII and IX, for hemophilia; and tissue plasminogen activator, for dissolving blood clots, are already being produced by other means and sold. Tissue plasminogen activator, for example, which is used primarily to help heart-attack victims, is currently derived from mammalian cells grown in laboratories. What makes the dairy-animal production method so appealing is that it is potentially much less expensive. To be competitive, however, we in transgenics must find a way to speed the process. Transgenic animals can pass on their special human gene to offspring, but creating a custom-made herd is a slow process—three years for goats and five for cows—because of the time required for gestation and growth to sexual maturity.

If the cloning of mammals were to become routine, however, new techniques could cut the current time frames in half. With cloning at their disposal, transgenics practitioners would be able to conjure as many animals as they liked from test tubes, rather than having to breed for two to three generations to create an appropriate herd. Since Ian Wilmut's pivotal achievement, a flurry of ground-breaking cloning experiments have been undertaken. As recently as late July, the Roslin Institute and PPL Therapeutics announced they had cloned a lamb possessing a human gene, using a technique that we at Genzyme Transgenics would consider performing on cows.

A tremendous momentum is building in the transgenics field, and each new success adds to the excitement. We are moving forward both in technical achievements and in acceptance by the pharmaceutical industry. One can therefore envision a world in which dairy factories would produce antibodies instead of cheese, and in which, from the udders of cows and other livestock, would come a host of medicines to fight cancer, AIDS, heart disease, and countless other serious ailments.

Proteins are the carpenters, the repairmen, the truck drivers of the body's internal world. They do the vital work that keeps systems running smoothly—enabling people to digest a hot dog, run a race, or fight off a cold. Enzymes, antibodies, and many hormones are proteins. Even the job of making proteins requires proteins. Thousands of kinds of such molecules—including familiar substances such as hemoglobin and insulin—course through the bloodstream at any given moment, acting as signals to the brain, the heart, the liver, and other organs.

Back in the late 1970s, when I was still a graduate student, biologists had just begun to explore the possibility of recreating the proteins found in the body. Indeed, much of the medical biotechnology industry has been based on the premise that just about any protein in the body can be turned into a useful drug. It seems intuitively obvious that those molecules have a purpose and that if they could be produced in large amounts and tested in the appropriate model, their use could be commercialized.

The body makes proteins according to instructions from its DNA—the complex code that contains all the information an organism needs to grow and function. DNA molecules are packaged in genes, and genes code for specific proteins. To artificially produce a protein, biotechnicians must first identify and isolate the associated gene. That process is known as cloning a gene, because it is achieved through the replication of bacteria, which reproduce by cloning themselves.

Some 100,000 genes are wound into the nucleus of a single human cell, and interspersed in and around them are a far greater number of DNA sequences whose functions are still largely unknown. Picking out a single gene from that chaos of information might at first seem as daunting as finding the proverbial needle in the haystack. But with the help of *Escherichia coli*, a prokaryotic bacterium found, among other places, in the human gut, the job can be done in a straightforward manner.

Imagine, for a moment, that the 100,000 or so human genes and the other

CLONING

DNA sequences around them—the human genome—tell a story; the gene you want is the equivalent of one word within that story, such as *cat*. Appealing at random intervals in the genome are certain recurring DNA sequences—the equivalent of words such as *and* and *the* in a piece of literature. Enzymes derived from bacteria can recognize those recurring sequences and sever the DNA at those sites. As a result, the genome is broken up into millions of pieces. Those fragments are genes—entire words—or more often, parts of genes—letters of the alphabet. If the gene you want has been shattered, it must be reassembled, much the way that the letters *c*, *a*, and *t* must be put together correctly to spell the word *cat*.

Workers insert each piece of DNA into an *E. coli* bacterium. The bacteria are spread on a plate, where they multiply; each genetically altered bacterium thereby reproduces the human gene or gene fragment it is carrying, in addition to its own DNA. The plate becomes covered with bacterial colonies, each of which contains many copies of one of the inserted fragments of the human genome. Biologists then probe for the fragments they want, using artificial, radioactively labeled DNA sequences that bind selectively to those fragments. Once the fragments have been identified, they can be reattached to form the gene of choice. Biologists can then make as many copies of the gene as they like—"clone" it—inside the ever-reproducing *E. coli*.

Many of the molecular biologists working in biotechnology, myself included, spent their graduate careers growing and manipulating *E. coli*, which had been used as a teaching tool for decades. That bacterium quickly became the workhorse of biotechnology. In addition to helping biologists clone genes, *E. coli*'s cellular machinery can be exploited to make human proteins, once human genes have been spliced into the bacteria's genetic material. Among the proteins already grown inside *E. coli* and placed on the market as medicines are alpha and gamma interferon, human growth hormone and insulin. At first *E. coli* was the organism of choice for creating all the new proteins of biotechnology. But because *E. coli*'s own proteins are relatively simple, the bacterium turned out not to be equipped to make all the substances that biotechnology companies would like to produce.

Every protein molecule is essentially a strand of amino acids, strung together like a pearl necklace. In longer, more complex proteins, the amino acids interact, forming, for example, disulfide bonds and causing the chain to bend, loop, and twist like a necklace dropped carelessly into a jewelry box. The result is a distinctive conformation, a unique shape that enables each protein to carry out its function in the body. Complex sugar molecules that can be critical to the functioning of the protein are also sometimes attached at key sites.

When biotechnology workers tried to use *E. coli* to make complex human proteins, they found that the strings of amino acids remained unfolded—without the disulfide bonds and sugar modifications crucial to their shape and function. So to produce those more complex proteins, biologists began growing mammalian cells in vats, a method known as tissue culturing. The cells are kept at body temperature and surrounded by nutrient-rich broth, to trick them into acting as if they were inside a living organism. Human genes are introduced into the cells, which make the associated protein and secrete it into the surrounding medium, from which it can be harvested.

Proteins made by the tissue-culture method include tissue plasminogen activator; DNAse, for cystic fibrosis; Factors VIII and IX; and erythropoietin (EPO), which is used to treat acute anemia. Tissue-culture proteins emerge from the vat folded into their proper configurations. But because of the limitations associated with artificially delivering oxygen and nutrients to suspended cells, cells in broth can be only about one-thousandth as concentrated as cells in an organism. As a result, the yield of proteins is relatively low, making the tissue-culture method extremely expensive.

A third and in fact the original method of protein production is to extract proteins from human blood plasma. Drugs containing antithrombin III, alpha-1-antitrypsin, Factors VIII and IX, gamma globulin, and human serum albumin are often made in that way. But blood products are not ideal because they can be contaminated by viruses such as hepatitis and HIV.

The fourth and most recent method of making useful human proteins relies on transgenic animals: animals that have a piece of foreign DNA stably integrated into their genome. To put the process into historical perspective, I should emphasize that the Genzyme Transgenics farm in Charlton is only two years old, and that it will be at least two more years before the first drug developed from transgenic animals could hit the market.

All animals and plants manufacture proteins naturally, and their ability to do so makes them potentially valuable as bioreactors. At least two companies have shown that it is possible to produce human proteins in plants: Agracetus, Inc., in Middleton, Wisconsin, and Biosource Technologies, Inc., in Vacaville, California. At Genzyme Transgenics, we have chosen dairy animals as protein sources because the mammary gland has evolved in part to secrete proteins efficiently.

A lactating female goat can produce three liters of milk a day, and the level of protein in milk is between 4 and 5 percent. We have been able to make transgenic goats that can each produce six kilograms of a pharmaceutical protein in one eight-month lactation period; that is more than a quarter of the annual supply needed around the world for such potent proteins as Factor VIII or interferon.

The mammary gland normally secretes simple protein molecules such as the caseins, which are small and lack disulfide bonds. But the gland can be induced to make properly folded complex proteins such as the ones found in human blood plasma. Milk-based protein production could be carried out with any mammal, including rabbits, pigs, and sheep. In fact, one of the first steps we follow in creating our proteins is to make transgenic mice and milk them using miniature teat cups attached to a human breast pump. We have produced more than thirty different proteins in the milk of mice, and so far we have been able to generate five of those proteins at high levels in goat's milk.

How are transgenic animals created? First we clone the gene for a desired human protein in *E. coli*. We then attach that human gene to the promoter region of a gene taken from a goat. Promoters are the parts of genes that control when and where those genes are expressed; we use promoters that direct gene expression in the mammary gland.

Under a microscope, the two-part gene construct is injected into a zygote, or one-celled embryo, from a goat; the altered embryos are then implanted into goats that serve as surrogate mothers. Then we wait: patience is a tough discipline in the protein-growth business. Only after a baby goat is born can we test whether the transgene—the human DNA—has integrated successfully into its genome. If it has, the newborn becomes a founder animal—the first to carry that particular human gene. Founder goats are grown to maturity and bred so that they can pass along the transgene to their offspring.

While the breeding goes forward, workers take milk from a founder female and begin trying to isolate the target protein. As it turns out, a unique purification method must be developed for each protein. The aim is to determine which chemicals and processes will separate out the protein on the basis of factors such as its size, weight, and charge.

What are the advantages of producing human proteins in transgenic animals? First of all, milk can be obtained easily: dairy animals are milked every day, and a goat produces milk for seven years. If a protein is successful and demand for it grows, production can be expanded relatively cheaply by breeding more transgenic animals. That beats the cost of building a new high-tech plant to house vats of sensitive tissue cultures. The largest expense associated with the dairy-animal method is that of extracting the proteins from the milk. Traditionally, half the cost of a protein is in its purification—tissue-culture producers, however, face the same challenge.

But obstacles to the dairy-animal method abound. The first hurdle is getting the zygotes to develop to term. Because of the various manipulations they must undergo, fewer than one in five of the microinjected goat embryos develop into live offspring. Furthermore, of the animals that are born, as few as one in twenty test positive for the transgene; for unknown reasons, human DNA often does not integrate successfully into the goat genome. Finally, of the small fraction of animals that end up carrying the transgene, not all express it by producing the desired protein in their milk. If the transgene comes to rest in the wrong region of the goat's chromosomes, the animal may produce only a small amount of the human protein—or none at all. For those reasons, to produce a single transgenic animal, workers must implant between 100 and 200 microinjected zygotes.

Other fundamental limitations also slow the process. In goats, the time from the microinjection of DNA until a founder animal is born is five months. Another eight months must pass before the animal can be bred, and not until still another five months later, after parturition, does the goat begin milking and producing the protein. Thus, for the founder animal alone, the time from injection to lactation is a year and a half.

But that is by no means the end of it. Between thirteen and twenty-six more months are needed to create a herd of transgenic goats from the founder—"scaling up," as it is called in the business. The best-case scenario occurs when both male and female founder animals are produced; the male can be widely mated while milk from the female is used to develop a purification method. The worst-case scenario is that the only founder animal is a female. Then breeders must hope she will quickly produce a transgenic male offspring so that that male can be bred with many females after reaching sexual maturity. But when a transgenic goat is bred, there is at most a fifty-fifty chance that the offspring will carry the transgene; therefore, the odds of producing a transgenic male are one in four at best.

Cloning would make it possible to generate limitless numbers of animals with the same genetic makeup. What would that do for the dairy-animal protein business? For one thing, it could help workers skip a generation—and all the time that entails—when making a transgenic herd. If the only founder animal is a female, workers could dispense with breeding and simply use her cells to generate many identical animals.

But the potential advantages of cloning could be much more far-reaching. The existing technique of microinjecting zygotes with a human DNA construct is an iffy business. It seriously compromises the viability of the zygote, and it leaves no way of knowing until after birth whether the transgene has integrated successfully. With cloning, by contrast, workers could begin by introducing human DNA into goat cells grown in culture. Methods are available for selecting only those cells that absorb the DNA construct into their genomes, and then growing limitless numbers of them in culture.

Consider the differences: In current practice, one waits nervously to see whether multiple implantations involving hundreds of carefully manipulated zygotes end up producing a viable founder animal. Even then, one waits at least one more generation to produce a transgenic herd. With cloning, on the other hand, workers could transplant the nuclei of cultured cells that are already known to be transgenic into oocytes (unfertilized eggs), implant the genetically altered oocytes into surrogate mothers, and—presto!—in one generation's time, an ideal herd of protein-producing animals.

Ideal, that is, because cloning would not only save time; it would also give transgenics practitioners a kind of control that is crucial but currently lacking. As I noted earlier, a transgene can end up randomly in one of many places on the genome of the zygote, and its placement can substantially affect how much of the protein the resultant animal will secrete in its milk. Workers today have no control over that effect. But if cloning became practical, workers could select only those body cells into which the transgene had achieved optimal integration. The cloned animals that resulted would be certain to be good protein producers.

Unfortunately, the process of cloning adult mammals is still extremely inefficient. Wilmut and his coworkers began with 277 sheep embryos before they ended up with the clone named Dolly. The goat herds we use at the Genzyme Transgenics farm are imported from New Zealand because that country is free of scrapie and other diseases. Hence, obtaining large numbers of oocytes would be difficult and costly. Until the odds that a cloned embryo will develop to term can be substantially improved, our microinjection system seems preferable.

For cows, however, the balance may tip the other way. Dairy cows would be extremely well-suited to making transgenic proteins because of their robust milk production. The time and expense involved has so far made cows impractical for most transgenics applications, however (although Genzyme Transgenics is hopeful about their promise and has been experimenting with using cows to produce human serum albumin, which can be used to restore normal blood pressure in shock and burn victims). Cows are so large that operating on them to retrieve zygotes, as we do with goats, is prohibitively difficult and expensive. In our existing cow-based program, we have an arrangement in which we get cows a few days before they are to be processed for meat; we breed them and then retrieve the zygotes at the slaughterhouse, a rather unwieldy arrangement. Cows cost much more than goats, and their long gestation period and growth cycle can often make them unsuitable for developing a drug in the competitive marketplace.

Cloning, however, would reduce those time frames to workable levels; and it would be substantially less expensive than making a transgenic cow is now. And because the raw materials of cloning are oocytes, not embryos, the process would enable us to bypass our complicated zygote-retrieval procedure and allow us simply to buy cow ovaries at the slaughterhouse. Cloning, therefore, could make transgenic cows a practical alternative—and potentially the next golden goose.

The trade in human proteins is a multibillion-dollar business. Like the frankincense and myrrh of old, many of these medicines are precious elixirs, and they are priced accordingly: Factor VIII, for instance, can sell for hundreds of thousands of dollars a gram. For many important proteins, the method of choice is still to extract them from plasma culled from the donated blood supply, say, at the Red Cross. Although somewhat risky because of the threat of tainted blood, the plasma method does not always lead to six-figure pricing: human serum albumin, for instance, can be had for as little as $4 a gram, because it occurs in blood plasma at high concentrations.

The tissue-culture method and, more recently, transgenics are challenging the plasma method for dominance. Practitioners of those newer techniques are attempting both to compete with the plasma method—making the same proteins at equal or lower cost—and also to expand the market by producing new proteins that are not found in blood.

Factors VIII and IX have been success stories for tissue culture. Because those proteins occur in blood plasma in low concentrations, they were already expensive. Practitioners of tissue culture had only to match those prices to corner part of the market, because their products are safer. But the tissue-culture method has not been able to make some other plasma-derived proteins, such as alpha-1-anti-trypsin and antithrombin III, cheaply enough to compete. That is why Genzyme Transgenics has focused on those proteins in its efforts to see whether transgenics can become commercially viable.

To get a sense of the economics involved, consider the cost of producing monoclonal antibodies, which are proteins not found in plasma. To make 100 kilograms annually of a monoclonal antibody, tissue-culture practitioners need to produce 170,000 liters of protein broth. That in turn demands a production plant with an 8,500-liter reactor capacity; according to Genzyme Transgenics estimates, such a plant would cost $50 million to build. The current price of monoclonal antibodies made in tissue culture ranges from $300 to $3,000 a gram.

By contrast, because of the high protein concentrations in milk, only 21,000 liters of transgenic milk are needed each year to make the same 100 kilograms of monoclonal antibody, and that demand can be met by just thirty-five transgenic goats. The animals could be generated and maintained in a farm setting complete with purification facilities for less than $10 million, our estimates say. With those lower production expenses, transgenics would be able to produce monoclonal antibodies for just $105 a gram—a third or less than the lowest tissue-culture price.

The time to market is critical, however, for many protein therapeutics. Practitioners of the tissue-culture technique have reported aggressive time lines of only a year from clone (of the gene) to clinic. Until now, there has been no way to speed up the process of making transgenic animals. But if the cloning of adult mammals—still an intriguing anomaly—ends up developing into a viable alternative with high success rates, the new field of transgenics promises to become much more than a drop in the (milk) bucket.

6   Brave New Rose

David Concar

IT'S 2020. YOU'RE LYING ON A LEMON~SCENTED LAWN. THE ROSES ARE BLUE.

Turning a rose blue. In an era when researchers can clone mammals and insert genes into plants to ward off crop-devouring insects, you would think this would be easy. But it isn't.

Ask Edwina Cornish. Years ago, this Australian biotechnologist and her colleagues began a quest to create in the lab what cannot be created by breeding. They founded a company, Florigene in Collingwood, Victoria. They raised money for the research. They cloned the gene that enables petunias to produce the blue pigment that roses lack. But when they inserted the gene into rose cells, the resulting flower was no bluer than, well, a rose.

Then there is the mysterious case of the mutant loblolly pine. Another dream of plant engineers is to create easy-to-pulp trees. For years, researchers believed the key in all species was an enzyme called cinnamyl alcohol dehydrogenase, or CAD.

Reprinted with permission from *New Scientist* 160, no. 2158 (October 31, 1998).

This, after all, was what the textbooks said all woody plants used to synthesize the lignin polymers that make cell walls sturdy and the extraction of cellulose costly.

But then last year, unexpectedly, Ronald Sederoff of North Carolina State University in Raleigh and his colleagues uncovered some mutant pines that broke the rules. The trees had a mutation that blocked all production of the CAD enzyme—yet they still made plenty of lignin. In pine trees at least, genetically manipulating levels of this enzyme would not dramatically help the pulp extractors.

You get the picture. Plant biochemistry is turning out to be more unpredictable—and harder to tame—than researchers had thought. Even the researchers say so. "In the early days it was easy to be optimistic," says Cornish. "We might have underestimated how long things would take and the complexity of the pathways we were trying to manipulate."

But here's the rub. Hard to tame does not mean impossible to tame. Slowly but surely, researchers like Cornish and Sederoff are getting to grips with the complexities of engineering plants. Slowly they are laying the foundations for a world where the initials "GM" will come to prefix far more than just genetically manipulated tomato puree and soya beans.

It might take five years, it might take twenty, but we will have genetically modified roses that are blue, along perhaps with GM geraniums that smell of roses, GM orchids that glow when they need watering, GM leylandii hedges that stop growing at a reasonable height, GM lawns that (almost) never need mowing, and GM bin liners made from plastics synthesized in plants. Not to mention GM newspapers and wallpaper.

If this sounds silly, think what has been achieved so far. Fifteen years ago, there was just one technique, based on the grown gall bacterium, for ferrying genes across the thick walls of plant cells. Now there are several, including two types of gun for propelling DNA into cells at high speed. A decade ago, researchers knew almost nothing about the genes that control the shapes, sizes, and flowering characteristics of plants. Now dozens of such genes have been identified and a project to sequence the entire genome of a flowering plant, a weed called *Arabidopsis thaliana*, is nearing completion.

Already efforts are well under way to engineer potatoes to double up as vaccines; to create transgenic "smart" plants that will use a fluorescent "SOS" protein to give farmers or growers early warning of drought or disease; to equip oilseed rape with bacterial genes for producing biodegradable plastic; and to engineer cotton plants to produce wrinkle-free fibers.

CLONING

FORESTS OF CLONES

One by one, even trees, which are notoriously tricky to grow from tissue cultures, as genetic engineering demands, are falling under the spell of biotechnology. As a result, timber and pulp will increasingly come from high-tech plantations where the trees are all clones, engineered to carry new genes for pest and disease resistance, and perhaps made sterile to prevent these transgenes escaping via pollen, says David Ellis of the BCRI Forest Biotechnology Centre in Vancouver.

Some tree plantations, mostly in the southern hemisphere, already consist of genetically identical trees, mostly produced the way gardeners and farmers have cloned plants for millennia—with vegetative cuttings. But as genetic engineering takes off, more and more forestry plantations will begin life as so many cloned tree embryos, frozen until they are needed and then cultivated in vast hydroponic vats. Granted, cloning can be labor intensive and genetic uniformity is not always desirable. But many growers are keen on it because it enables them to raise the quality of all their trees to that of the best. The availability of "elite" genetically manipulated trees will make them even keener.

"Biotechnology will accelerate the trend toward clonal forestry," predicts Martin Maunders of Cambridge-based biotech company ATS.

One "elite" trait would be easy-to-extract pulp. Ellis points out that despite last year's mutant pine surprise, the synthesis of lignin in other commercial tree species is actually "very well characterized." "We know and have isolated every gene in the biosynthetic path," says Ellis. And in eucalyptus and poplar trees at least, engineering levels of the CAD enzyme does make pulp easier to extract.

Researchers elsewhere are experimenting with genes that may boost the growth of trees during winter months or curb the height of fruit trees so they take up less space and their fruit is easier to harvest. Mini cherry trees small enough for the tiniest city garden could be just a few years away. In labs like Sederoff's, meanwhile, efforts are under way to identify genes that affect wood strength and density. Where will it all lead? "A short, fat, fast-growing tree" might be the thing of the future, says Ellis, only half in jest. "With no taper so you don't waste space on the logging trucks."

And with technicolored timber perhaps. For there's nothing about the biology of plant pigments that means grass has to be green or that wood has to be a yellowy brown. Polka dot button holes to match your tie or scarf are some way off, but already a couple of transgenic carnations that are mauve rather than the usual pink, yellow, white, or red are being sold by florists in Australia, Japan, and the United States.

The carnations owe their strange hue to the pigment gene Cornish and her colleagues cloned from petunias—the one that has so far failed to turn roses blue. Why the gene works in carnations (up to a point) but not in roses isn't entirely clear. But the researchers suspect petal acidity is a major factor. The gene encodes an enzyme needed to synthesize blue pigment molecules called delphinidins, which are lacking in both roses and carnations. The problem for roses is that these molecules are only blue at high pH, and the cellular cavities, or vacuoles, that hold petal pigments in roses are acidic.

To solve the problem, Cornish and her colleagues are pinning their hopes on one of two options—finding a conventional rose variety that is less acidic, or cloning the genes that control petal pH so that they can alter conditions in the vacuoles. Even then, there remains a risk that the rose's natural pigment molecules, the red cyanadins and orangy perlagonidins, will drown out the added blue.

One reason why turning grass blue or red might be easier than it sounds is that any biochemical changes might only need to be skin deep. For instance, genetic engineers could use a pigment gene hooked up to a piece of DNA that keeps the gene switched off in all but the outer layer of cells.

Another approach might be to make use of silent and unused pigment genes. After all, the green stems and leaves of ornamental flowers have the same genes as the petals. "The genes are there," explains Cornish, "but they are expressed in the flowers, not the leaves." In theory, genetic engineers could rouse these pigment genes from their slumber, producing leaves and stems awash with floral pigments.

So, when blue roses do finally begin to emerge from labs, perhaps some of them will have the chance to express their native pinkness in their leaves.

Some might also have the chance to smell of lemons. "Some people find sweet roses overwhelming, and most cut roses have almost no odor at all," says Michael Dobres.

Two years ago, in Philadelphia, Dobres helped found a company called NovaFlora that aims to remedy this sorry state of affairs. One of their projects involves inserting a gene into roses that would enable their petals to produce lemon fragrance molecules. The gene encodes an enzyme called limonene synthase, which citrus plants use to synthesize scent molecules known as monoterpenes. The researchers have already given the gene to petunias and are waiting for their first crop of what they hope will be a lemon-scented transgenic flower.

Limonene synthase is only one way to perk up scentless plants. "There are hundreds of different monoterpenes out there, synthesized by different enzymes," says Dobres. Not to mention two other major types of plant fragrance molecule. In future, predicts Dobres, genetic engineers will be able to create finely tuned fragrances to order in almost any plant. Among the many possibilities would be lemon-scented golf courses and GM camomile lawns that are much easier to maintain than the traditional kind. And as for the idea of Calvin Klein-scented GM roses, "That would be dynamite," says Dobres. "That's something we definitely aspire to."

Of course, achieving all this won't be easy. The scent molecules that transgenic plants make will be produced in vain if they remain trapped inside their tissues. One reason many commercial cut flowers are so odorless in the first place is that breeders select for tough petals with waxy coats. Then again, perhaps genetic engineering could be used to make these coats permeable to scent molecules.

It can certainly be used to alter the shape, form, and number of flowers that a plant produces. Knowledge of the gene code which specifies the physical arrangement of a flower's sepals, petals, stamens, and carpels is so advanced that it is already possible to design "fantasy flowers" that have any of these organs in any position in the flower. And genetic engineers can also alter when a plant flowers.

At the University of Leicester, Garry Whitelam and his colleagues have engineered asters so that they flower in the middle of winter, not just in summer. Growing conventional cut flowers in greenhouses in winter is expensive because of the extra lighting needed to make them flower. In a bid to cut costs, the researchers manipulated an aster gene so that it would produce higher than normal levels of a phytochrome protein that enables plants to sense changes in daylength. The GM asters required only six hours of daylight to flower compared with the usual fourteen.

And when it comes to manipulating the sex lives of plants, this is only the tip of the iceberg, thanks in no small part to *Arabidopsis thaliana*. In less than two decades this unprepossessing weed with white flowers has risen from obscurity to become the megastar of plant science. The attraction for researchers is that it has an unusually small genome and grows to maturity in just six weeks. And in the 1980s they decided to make it their fruit fly—the model organism they would mine for important genes involved in plant development.

In the past few years, such genes have been tumbling out of *Arabidopsis* labs, turning the heads of plant biotechnologists everywhere. Three years ago, for instance, Detlef Weigel of the Salk Institute in La Jolla and Ove Nilsson at the Swedish University of Agricultural Sciences in Umeå identified two genes in *Arabidopsis* that act as master switches for triggering flower formation at the ends of shoots. When the researchers engineered *Arabidopsis* so that the genes would be active all over the plant, every shoot produced a flower. And when they inserted one of the two genes, *leafy*, into aspen, a tree that normally takes up to two decades to flower was fertile after two months. A spectacular result given that slow sexual development is the bugbear of tree breeding.

DEATH SIGNAL

Other researchers are exploring ways of using *Arabidopsis* genes to do the exact opposite—prevent flowering. And not just to prevent transgenes spreading into wild relatives. For annual crops such as lettuce and potato plants, flowering is a prelude to death. It sends a signal to the leaves telling them to shut down photosynthesis. Blocking that signal might mean farmers could grow the crops for longer and perhaps get bigger yields because the plants would no longer need to invest resources in making flowers.

Nobody has engineered crops this way yet, but the discovery of an *Arabidopsis* gene called *Frigida* could encourage researchers to try. In the weed, the gene seems to function as its name suggests—to prevent flowering, or at least to delay it until winter is over. "It would be nice to stick the gene into sugar beet and see what happens," says Caroline Dean, at the John Innes Centre in Norwich.

In future, farmers and growers may even use chemical sprays to make their genetically engineered plants flower on cue. Earlier this year, Brian Tomsett and Mark Caddick at the University of Liverpool used an alcohol-sensitive gene from a fungi to make the activities of plant genes controllable from the outside. Simply drenching the roots of the engineered plant with alcohol was sufficient to switch on a gene that stunted growth.

And why stop there? Why not manipulate plants so that they can change, on cue, their color or fragrance? Why not engineer fast-growing hedges whose growth can be "switched off" once they reach the required height?

Why not...create a GM world?

7

Send in the Clones...
Don't Bother, They're Here[1]

B. Rollin

I

I n February of 1997, Scottish researchers announced that they had cloned a sheep, appropriately named Dolly, from a cell derived from the mammary tissue of an adult sheep. Dolly appeared unremarkable, and was phenotypically identical to any other sheep. She was, however, genetically identical to the sheep from whose mammary tissue she had been created by introducing the DNA from that cell into an unfertilized sheep egg from which all DNA had been removed. This egg was then inserted into a surrogate mother, to whom Dolly was born in the normal manner.

Although mammalian clones had been created before, the mechanism for their production had been the splitting of developing embryos. Dolly, on the other

From *Journal of Agriculture and Environmental Ethics* 10, no. 1 (1997): 25–40. Copyright © 1997 Kluwer Academic Publishers. Reprinted with kind permission from Kluwer Academic Publishers and the author.

hand, was dramatic proof of what had long been a theoretical possibility in animals and an actuality in plants—the creation of an organism by use of the genetic material contained in a somatic or bodily cell, rather than a reproductive or germ cell. This in turn could occur by virtue of the fact that all nucleated bodily cells are *totipotent*, i.e., contain the entire genetic blueprint for the whole organism. The birth of Dolly demonstrated that one could, in principle, create new organisms from any bodily cell of any organism.

Although the cloning of Dolly was enthusiastically acclaimed by most biologists, such was not the case in society in general. Theologians predictably condemned humans usurping the role of God and call for a ban on such research. But the general public too expressed fear, revulsion, and "ethical concern," prompting the British government to cut off additional funding to related research. According to a CNN/*Time* magazine survey of 1,005 adult Americans released one week after Dolly's birth announcement, "most Americans think it is morally unacceptable to clone either animals or humans, and that new cloning techniques will create more problems than they solve" (CNN/*Time* 1997). Fifty-six percent said they would not eat cloned animals, while more than half said they would eat cloned plants, and two-thirds said that the federal government should regulate cloning of animals. Even greater apprehension greeted the possibility of cloning humans. Sixty-nine percent said that they are "scared" by the prospect of cloning humans, and 89 percent said that such cloning would be "morally unacceptable." Seventy-five percent said that cloning is "against God's will," and 29 percent were sufficiently troubled by the new technique for reproducing life that they would be willing to participate in a demonstration against cloning humans. Only 7 percent said that they would allow themselves to be cloned if given the chance.

Media coverage of Dolly was considerable, and the story was front-page news for days. As one who writes in bioethics, I received five phone calls from newspapers and broadcasters in a single day, as well as numerous calls and visits from lay people—and scientists—worried about the "ethics" and "ethical issues" occasioned by cloning.

Despite the pervasive outcry about the "ethical issues" and "moral unacceptability" of cloning, there was little rational articulation or clarification of what these issues in fact are, or why cloning is morally unacceptable. Instead, as just mentioned, one heard a great deal about cloning "violating God's will," being "against nature," or attesting to the power of margarine advertising to shape Western thought, it "not being nice to fool Mother Nature." Further, much of the concern was based on factual misinformation. For example, the pope affirmed that cloning was wrong because all creatures had the right to a natural birth, thereby both forgetting that Dolly did have a natural birth and that in any case he was begging the question had she not had one. Others, forgetting that Dolly had been born some seven months before her birth was announced, assumed that cloning gave rise to fully formed adults. Still others assumed in their critique that, through cloning, one could produce indefinite numbers of *literally identical* individuals, not just *genetically identical* individuals—as we shall see, the two concepts are far from identical.

In our ensuing discussion, we shall examine both the alleged and genuine socioethical issues associated with the arrival of this new technology. It is essential that philosophers (or someone) undertake this task, to avoid two possible socially and scientifically deleterious consequences. First of all, issues not truly moral may be mistakenly characterized as ethical issues, and restrictive social policies instituted which retard scientific progress based on these fallaciously designated concerns. Not everything that upsets people—even large numbers of people—is a moral issue. For example, were we to ban cloning on the basis of people's belief that it was immoral because it violates God's law, we would be responding to a theological concern, not a moral one in a secular society whose ethics is (or ought to be) divorced from religion. Similarly, were we to ban it because many people see it as "unnatural," without a clear account both of what "natural" and "unnatural" means and why the unnatural is immoral, we would again be confusing revulsion with morality—few of us would advocate banning aeronautical research on the grounds that flying is unnatural for human beings. There are in any case natural clones—they are of course identical twins!

II

These reflections lead me to a major concern about the cloning of Dolly. It is not the cloning per se that I find problematic. It is rather our total social-ethical unpreparedness to deal with that event morally and rationally. It is not as if this lack of preparedness is justified. The theoretical possibility of cloning has been with us for at least three decades, actual cloning (via splitting embryos) for over a decade. Even more important, the scientists who achieved Dolly's cloning have obviously been working for some time to accomplish this goal. Why were they not, during the course of their research, educating the general public and the mass media as to the nature of cloning, and inaugurating the ethical and social discussion logically necessitated by the advent of this technology? More generally, why wasn't the scientific community, or at least the relevant part of the scientific community, preparing society to take intelligent positions on the science and technology they were working to develop? In my view the ethical issues occasioned by scientific or technological progress—even biotechnology—are no more inherently difficult to resolve than more familiar social-ethical issues; health management, capital punishment, affirmative action. They appear more vexatious only because we don't understand the relevant empirical background, and have correlatively not even mapped the ethical concerns, and lack the basis for rational discussion. In short, in science and technology our ethical knowledge and sophistication rarely if ever keep pace with scientific progress.

This is socially unacceptable if we are committed to a rational democratic society, as opposed to Plato's caricature of democracy as a team of horses pulling in different directions. Any new and powerful technology—be it the automobile, nuclear power, or cloning—must be socially regulated or society must decide not to regulate it. If we are to be rationally democratic, then the principles behind regulation or the lack thereof must reflect a consensus, not special interests or rule by "experts," who typically have their own agenda. Yet presuppositional to such a broadly based consensus is social understanding of the technology and its implications. Failure to educate the population as a whole must lead to irrational, fear-based, knee-jerk decision making founded in rumor, not in knowledge. Thus it is in the interest of the scientific-technical community to educate society as a whole as to the nature of new advances, so that people can dissect out the ethical issues, the risks, and the areas requiring regulation and oversight. By failing to do so, science and technology puts itself at risk, as public reaction to BST and cloning shows. We have already seen that the public's response to Dolly has led to the curtailment of British governmental funding for such research.

If both proper functioning of the kind of society we value and rational self-interest militate in favor of the scientific community's undertaking the education of the public on the nature, risks, and ethical issues of emerging technologies and scientific breakthroughs, why is such education so rarely undertaken, let alone successfully? In the first place, because there is no readily apparent vehicle for such education. Second, because scientists are generally ill at ease talking to lay people, since the latter do not share their language, values, or knowledge base. Third, it is because for much of the twentieth century science has been captured by an ideology that affirms, among other things, that science is "value-free" and thus has no truck with ethics (Rollin 1989)—a corollary of this ideology is that it is society's job to identify, articulate, and deal with ethical issues emerging from science and technology. Ironically, while following this ideology, the scientific community at the same time deplores the way society emotionally and ignorantly articulates and expresses its moral concerns and regulates and constrains science on the basis of that articulation!

Fortunately, the hold of this once ubiquitous ideology is being eroded by

CLONING

numerous factors—public law mandating that scientists engage the ethical issues emerging from research on human and animal subjects; increasingly vocal pressure from subgroups of society such as women, homosexuals, and animal advocates forcing the research community to deal with ethical issues such as medical research being done largely on male subjects, or the medical community's failure to acknowledge pain in animals or human neonates. Fourth, the education of scientists has been liberalized. For example, the National Institutes of Health now mandate education in science and ethics for at least some young scientists in training.

In any case, biological science and technology such as genetic engineering, advancing as they do with great rapidity, absolutely require an educated public who have had reflective time based on input from scientists in the field to take a thoughtful position on these advances and their regulation. In the absence of such a process, the ethical void is filled by doomsayers, sensationalists, and people with an axe to grind, such as theologians. It is for these reasons that in my book on the ethical and social issues occasioned by genetic engineering, I have argued for local citizen panel review of all genetic engineering projects, so that citizens feel as if they both have a voice and a stake in regulating, preventing, or advancing the new science and technology (Rollin 1995b). The same principle is easily exportable to cloning. Whatever society decides to do with this new science, it should be done on the basis of knowledge, understanding, and dialogue, not fear, misunderstanding, and emotion.

III

In my discussion of the genetic engineering of animals in *The Frankenstein Syndrome* (Rollin 1995b), I argued that the possible socioethical issues associated with that technology fall into three categories. These categories, which I shall briefly review, are relevant to our discussion, as they can be applied, mutatis mutandis, to cloning. The first are essentially spurious issues of the sort we have already described—fear of the unknown and the misunderstood, ignorance, aesthetic revulsion, religious distaste, all sloppily categorized as ethical concerns. The second are issues of risk associated with genetic engineering—risk of environmental despoliation by virtue of release of genetically engineered organisms into unintended environments; risk of new disease growing out of changing the animals in both immunological and non-immunological ways and unwittingly selecting for new pathogens dangerous to humans or other animals; risk of genetically engineered animals such as the SCID mouse, designed to be able to be infected with the AIDS virus, either infecting humans or having the endogenous mouse viruses interact with the AIDS virus to produce pathogens with unpredictable characteristics. One should note that controlling the risks is not in itself a matter of ethical debate—all parties to the discussion, be they genetic engineers or ordinary citizens, have an interest in preventing catastrophes. The moral issue that arises devolves around how much risk one ought to be willing to assume in the face of what possible benefits. While scientists may deploy one set of values in weighing heavily the benefits of creating the SCID mouse, and correlatively minimize the importance of the risks and emphasize their ability to control them, ordinary citizens may well be unwilling either to suffer any risks in order to create a mouse model for AIDS, or to place their own perceived security in the hands of scientists' assurances about the certainty of containment.

Finally, the last and most difficult moral issues in the genetic engineering of animals arise out of the possibility of causing harm, pain, or suffering to the engineered animals. These issues are most difficult because it is the animals that will suffer, not humans, and one must be prepared to weigh human benefit against costs to animals. Given that industrialized agriculture has shown little concern for any aspects of animal welfare other than those that impair efficiency or productivity of the animals (Rollin 1995a), we cannot rule out the accidental or intentional creation of and propagation of suffering animals for purposes of commercial agriculture. By the same token, our new-found ability to create animal models of grave human genetic disease again suggests that such suffering animals will be created as a tool in the fight against human disease, even though the cost is uncontrollable animal suffering (Rollin 1995b). In sum, then, there are spurious ethical issues; issues of who should bear the risks of genetic engineering, what risks are possible, and who should decide what risks are worth taking for what benefits; and issues growing out of harming animals for human benefits in significant ways and in significant numbers.

This typology is a good template for approaching the cloning of animals. We have already discussed the spurious issues—these may be dismissed from rational ethical dialogue, though most assuredly not from political dialogue. What remains, then, are issues of risk and issues of animal welfare.

Are there any risks associated with cloning animals? The most significant risk seems to me to arise from the potential use of cloning to narrow the gene pool of animals, particularly in agriculture. With the advent in the mid-twentieth century of an agriculture based in a business model, and emphasizing efficiency and productivity rather than husbandry and way of life as supreme values for agriculture, it is now evident that sustainability has suffered at the hands of productivity—we have sacrificed water quality to pesticides, herbicides, and animal wastes; soil

CLONING

quality to sodbusting and high tillage; energy resources to production; air quality to efficiency (as in swine barns and chicken houses), rural ways of life and small farms to large, industrialized production techniques. What is less recognized but equally significant is that we have also sacrificed genetic diversity on the same altar.

A lecture I once attended by one of the founders of battery cage systems for laying hens provides an excellent example of how this works. He explained that, with the rise of highly mechanized egg factories, the only trait valued in chickens was high production, i.e., numbers of sellable eggs laid. Laying-hen genetics focused with great skill and success on productivity. Inevitably, the production horse race was won by a few strains of chicken, with other traits deemed of lesser significance. Given the efficiency of artificial selection and rapid generational turnover in chickens, the laying chicken genome grew significantly narrower. Thus today's laying hens are far more genetically uniform than those extant in the 1930s. In fact, said the speaker, such selection has so significantly narrowed the gene pool that, had he known this consequence, he would never have developed these systems!

Why not? Because the narrowing of the gene pool in essence involves, pardon the execrable pun, putting all our eggs in one basket, and reduces the potential of the species to respond to challenges from the environment. Given the advent of a new pathogen or other dramatic changes, the laying hens could all be decimated or even permanently destroyed because of our inability to manage the pathogen. The presence of genetically diverse chickens, on the other hand, increases the likelihood of finding some strains of animals able to weather the challenge.

Cloning will almost inevitably augment modern agriculture's tendency towards *monoculture*, i.e., cultivation and propagation only of genomes that promise, or deliver, maximal productivity at the expense of genetic diversity. Thus, for example, given a highly productive dairy cow, there will be a strong, and inevitable tendency for dairy farmers to clone her, and stock one's herd with such clones. And such cloning could surely accelerate monoculture in all branches of animal agriculture. Cloning could also accelerate our faddish tendency to proliferate what we think are exemplary animals, rather than what we might really need. For example, very high production milk cows for which we have selected, have very short productive lives; very lean pigs are highly responsive to stress, etc.

At the moment, agriculture's only safety net against ravaged monocultures are hobby fanciers and breeders. Although egg producers disdain all but productive strains, chicken fanciers, hobby breeders, and showmen perpetuate many exotic strains of chickens. Given a catastrophe, it would surely be difficult to diversity commercial flocks beginning with hobby animals as seed stock, but at least genetic diversity has been preserved.

One suggestion that has been made to prevent loss of genetic diversity is the establishment of a "gene library," wherein one could preserve the DNA of animals no longer commercially useful, and also prevent the extinction of endangered species. A major problem with this concept is that the genome in question is, as it were, temporally frozen. Whereas chickens in the world would be adapting to changing conditions, albeit probably slowly and imperceptibly, the stored genome would not. Thus the genome stored in 1997 might not have kept up with environmental changes that have occurred by the time we need to resurrect these chickens in 2997. A second, minor problem arises from the probable slowness of the response time required to resurrect the genome so that it is functionally operative in the requisite environment.

There is also a deeper problem with gene libraries. Such repositories for DNA might well create a mind-set that was cavalier about the disappearance of phenotypic instances of endangered organisms. If people believe that we have the species in a state of, as it were, resurrectable suspended animation, they might well cease to worry about extinction. This is problematic on a variety of levels. First of all, there is a huge aesthetic (and metaphysical and perhaps ethical) gap between preserving white tigers on the tundra in Siberia and preserving them in zoos, let alone in blueprint form in the test tubes. Furthermore the tiger does not express its full *telos*, its tigerness, independently of the environment to which it is adapted. To preserve the tiger while allowing its *umwelt* to disappear is to create the biological (and moral) equivalent of stateless, displaced persons. The prospect of white tigers with no place to be in the world but casinos in Las Vegas, wild animal parks, or zoos is not an acceptable solution to the problem of endangered species. Indeed, one could argue that such a solution is little better than extinction, since one has de facto extinguished the tiger as a form of life, though not as a life-form.

Other than the encouragement of the deplorable tendency towards monoculture, I can see no risks associated with cloning. The question now arises as to whether cloning is likely to have a negative effect upon the well-being of cloned animals.

The first such possible consequence arises out of the possibility that cloning per se can have unexpected and deleterious effects on the animals. Although one is putatively creating an organism that ought to end up indistinguishable from a naturally derived animal, it is conceivable that the process of cloning could itself have deleterious effects that emerge at some stage in the life of the organism. This phenomenon has already been manifested in cattle clones created by splitting embryos by nuclear transfer. According to veterinarians working with these animals, they have been oversized and thus difficult to birth, had difficulty surviving, and have also been behaviorally retarded, requiring a good deal more care at birth than normal calves (Garry 1996). (Indeed there seem to be problems in noncloned animals created by in vitro fertilization.) The cause of this is not known, and it is quite possible that clones like Dolly could "crash" later in life in virtue of some unknown mechanism. At this point, there is no evidence for this concern—it is an empirically testable possibility that will be verified or falsified as our experience with cloned animals develops. If it turns out that there are in fact unanticipated welfare problems for animals that are cloned, this should and likely will abort the technology until the problems are solved. There is a precedent here from the unanticipated consequences of inserting the human growth hormone gene into pigs. While these transgenic animals were leaner and faster growing as expected, they also displayed totally unexpected and disastrous effects, including lameness, synovitis, cardiac disease, ulcers, reproductive problems, and kidney and liver problems (Pursel 1989). In sum, cloning should not create pain and suffering in animals, and our emerging social ethic for animals clearly has the minimizing of animal suffering as its major thrust (Rollin 1995b).

There is a more subtle sense in which cloning can conceivably create problems for animal welfare and thus give rise to genuine social moral concerns. It could be argued that cloning feeds directly into a view of animals that has led Western societies to demand a higher moral status for animals. I am referring to what my colleague, Michael Losonsky, calls the *commodification* of animals, or of life. As I have discussed at length elsewhere, for most of human history the major use of animals by humans was agriculture—food, fiber, locomotion, and power—and the essence of agriculture was *husbandry*, care and stewardship. Etymologically derived from the Old Norse hus/bond (=bonded to the house), husbandry meant putting animals into an environment for which they were optimally suited by nature, and augmenting their natural ability to survive and thrive with additional food, shelter, medical attention, protection from predation, etc. Temple Grandin calls this the "ancient contract," wherein both humans and animals were better off in virtue of that relationship. In husbandry-based agriculture, no rational human would hurt their animals in any prolonged or extended way, for one would destroy the animal's productivity and ultimately harm oneself. Husbandry agriculture was thus about putting square pegs in square holes, round pegs in round holes, and creating as little friction as possible while doing so.

In the mid-twentieth century, however, humans broke the ancient contract with animals through emerging technology. We could now put animals into environments for which they were not biologically suited, yet keep them productive through technological "fixes" such as antibiotics, vaccines, and hormones. Although the animals surely suffered in such environments, their productivity (or more accurately, the productivity of the operation) was essentially unimpaired. Instead of worrying about the whole animal, we could ignore needs unrelated to productivity. With technological sanders, we could fit square pegs into round holes.

Rollin: Send in the Clones . . . Don't Bother, They're Here

Animal husbandry became animal science, defined as the application of industrial methods to the production of animals. In my view, this is the main factor that has called forth ever-increasing and international social concern about animal treatment. But whether I am correct or not, this modern approach inexorably leads to viewing animals as production machines, rather than as living things with lives that matter to them. Instead of husbandrymen, industrialized agriculture utilizes businessmen and managers, people for whom the financial reward is a significantly higher value than the way of life so central to traditional agriculturalists.

In such an agriculture, animals are products, pounds of pork, eggs per cage, commodities. The ability to clone them, one might argue, augments and reinforces this view. After all cloned animals are manufactured, and, like cars or soup cans coming off an assembly line, are "identical."

I have some sympathy with this concern, the same concern that informs animal advocates' vigorous opposition to patenting animals. But the issue here is far more basic than cloning—it is the industrialization of animal agriculture and the loss of the ethic of husbandry. On traditional hog farms, for example, sows had *names* and received individual attention. In today's huge production units, they do not. Cloning per se is perhaps a reflection of this industrialization, but there is no necessary connection between the two. After all, one can imagine a strongly husbandry-based agriculturalist caring a great deal for his herd of cloned pigs. Just because cloning has emerged from a questionable mind-set, does not mean that it could not thrive in a highly morally acceptable agriculture. Admittedly cloning is far likelier to be developed and employed in an industrial mind-set, but it does not follow that there could not be a use for it in a softer, more morally concerned agriculture. Just because cloning is a spin-off from industrialized agriculture does not mean that it is conceptually incongruous with sustainable husbandry. Western ranchers—the last husbandry agriculturalists—will continue to provide husbandry for their animals whether they are produced by AI, cloning, or natural breeding; after all, they are still animals under our care.

It could perhaps be claimed that cloning will accelerate public apathy about farm animal treatment, based on the psychological fact that the more of something we encounter, the more we see the units as interchangeable, the less we care about each unit. I doubt that cloning will worsen the situation we already have—though it won't improve it either. Phenotypically all noncloned laying hens, all broilers, all black cattle, all white sheep or pink pigs or white laboratory rodents look alike to the average urban citizen. This has not served to diminish social concern about their treatment—concern for the treatment of these uniform animals in the laboratory or on the farm has continued to grow, not diminish, as evidenced by legislation in the UK, Sweden, and the EU.

CLONING

I can imagine people wishing to clone their beloved dog or cat or horse—companion animals—even as some people taxidermy or freeze-dry their pets. I see such a use of cloning as unproblematic and self-limiting. It is not problematic because we will never have monoculture of pets—there are too many different preferences. Cloning would certainly not cause as many welfare problems as breed standards do today, wherein we perpetuate hundreds of genetic diseases in dogs by our aesthetic predilections. Indeed, I see no welfare problems here unless cloning has the adverse effects we mentioned earlier. Such cloning, I believe, would be self-limiting, when people realized that the clone is not their beloved Fifi and may be very different. At that point, people will realize that they may as well just get another puppy from Fifi's bloodline, if she is purebred.

III

The major concern expressed about the cloning of Dolly seems to derive from a "now we're headed down the road to perdition," slippery slope sort of argument. Natural barriers have been breached; it is simply a matter of time before we will be cloning *humans*! It is clear that the notion of cloning humans is shocking, perhaps repulsive, to significant numbers of citizens. The key question for us, however, is what ethical issues it raises. Is the revulsion to be cashed out in genuinely moral terms, or is it simply the shock of [the] unfamiliar, the counterintuitive, the scary?

One of the first issues requiring clarification is a metaphysical one. Many people's negative response to cloning of humans derives from their belief that one is creating identical copies of the same individual; multiple instantiations of what is, at root, the same person. While questions of personal identity are notoriously difficult for philosophers, the issue in this case is fortunately much clearer. For while clones may be genetically identical, they are not literally the same person, any more than natural clones—identical twins—are the same person. Science-fiction scenarios notwithstanding, clones will not be thinking the same thoughts at the same moment, nor will a clone of you be you.

Two individuals having the same genetic pattern are analogous to two houses being built from the same blueprint. In ordinary language we may talk about such houses being the same house, as in "Mrs. Smith and Mrs. Jones bought the same house in that development." But few people, on hearing that phrase, believe that the Smiths and the Joneses will be showering together. Just as the two houses will differ in many respects, so will a clone of an individual the clone is derived from, or two derived from the same individual.

Rollin: Send in the Clones . . . Don't Bother, They're Here

Let us suppose that I am cloned; something my colleagues have frequently suggested, given my type A personality, my tendency to overcommit, and my cavalier disregard for calendars, which has led me on occasion to promise to be in two places at once. Tired of these embarrassing scenarios, I decide to create a clone of myself and, as tomorrow's vernacular might have it, "do a Dolly." Let us further assume that I have had my genetic material inserted into an egg, and it is happily developing. A variety of forces will influence gene expression, including random factors and the prenatal environment. Thus, which genes express, and to what extent, will vary from clone to clone. Thus variation will be assured from the earliest stages of development. When George Seidel at Colorado State University cloned the first calves from a split embryo, the differences between them were phenotypically evident. For example, they were Holsteins (the familiar black and white dairy cows), and the pattern of black and white was markedly different among the clones.

So from birth and even before, all clones will be different from each other and different from the parent organism. Differentiation will continue to occur as little Bernie grows in virtue of environmental or natural considerations. For example, whereas I was raised in New York, Bernie Jr. will be raised in Colorado. Whereas I was raised with one parent and a grandmother, Jr. will be raised by two parents, one of whom is me. If my wife hates the idea of a cloned Bernie, that will certainly shape the child's emotional life, as my wife's cooking will shape his physical development. Although my mother discouraged my participation in sports, I will bestow a set of weights on Jr. while he is in the crib. And so on. In fact, my identical twin—if I had one—is far more likely to be similar to me than my artificial clone, since the twin and I experienced more or less the same environment. Yet that situation did not make us the same person.

We may thus dismiss the favored media examples of making dozens of Hitlers or Michael Jordans via cloning—the Hitlers almost certainly would not all emerge as dictators, the Jordans probably would not all end up star basketball players. Were we to start one of the little Adolfs on art lessons, he may well end up an artist. Were we to raise one in a Jewish household, he might well not become anti-Semitic. And so on. Similarly, the Michael Jordan clones might well not even be interested in basketball and instead might excel in some other sport or, depending on how they are raised, might have no interest whatsoever in sports.

As near as anyone can tell, our destinies are neither wholly determined by our genes (contrary to genetic determinists) nor wholly by our environment (contrary to behaviorists and Marxists), but rather by complex interplay between the two. Indeed, if we wished to create a whole team of superb basketball players, we would probably do better, rather than cloning Michael Jordan, to pay him well to impregnate the world's finest female basketball player or players and pay all of them to raise the progeny in an environment highly focused on basketball.

Thus clones would not be identical to each other or to the parent. Further, given environmental influences, age differences, differences in hair and clothing styles, no one could recognize my clone as a clone, rather than as my normally derived progeny or grandson, were I older when he was cloned. Assuming no untoward effect of cloning per se, as we discussed earlier, neither he nor anyone else would ever need to know that he was clonally derived, anymore than children need to know that they were adopted.

In fact, it is hard to see what ethical issue is raised even if clones were identical. Do quadruplets or quintuplets upset the moral order? We would probably disapprove of someone having a dozen clones made at once, but we would probably be equally concerned about someone impregnating twelve wives at once. One possible moral concern is that via cloning single parents could easily have children who would lose the benefit of two parents. But this is, of course, already the case, given broken homes, artificial insemination, adoption, children born out of wedlock, and so on.

In sum, I can as yet see no moral problem with cloning a child except perhaps that the child might be stigmatized or marginalized. This in turn could be prevented by careful maintenance of confidentially or, alternatively, by large numbers of people electing to reproduce by cloning. When I was a child, it was shameful to be a child of divorced or separated parents. Such, of course, is no longer the case.

Another moral problem that has been raised is the fear that people will create clones of themselves for "spare parts," for example, a heart. This is, of course, farfetched. For one thing, one would need to wait until the child reached adulthood to use the parts. For another, animals whose organs have been genetically engineered to be compatible with the human immune system are close to realization. (This, of course, raises a separate moral issue of animal exploitation.) Third, such use of a human is clearly covered and forbidden by our current social ethic—it would be premeditated murder. Society would not allow this any more than one is allowed to sell one's children for their organs. Fourth, it would be enormously inefficient to raise people for spare parts. What seems much more likely to me is the development of a technology whereby one can clone "spare organs" from the relevant tissue, rather than growing and discarding the entire person. Such organ cloning seems in principle possible.

If, as I have suggested thus far, there is nothing inherently wrong with cloning a person, why do we find such a visceral negative response to it among ordinary people? Some of that reaction, I believe, can be explained by reference to deeply inculcated theological biases and a correlative sense of human *hubris*, chutzpah, man overstepping his bounds, transgressing what is forbidden—a main theme in Western literature—vide Daedalus and Icarus, the Golem, Frankenstein, the Rabbinical tradition about Kabalistic knowledge, the Sorcerer's apprentice. We have heard similar rumblings at every major technological advance—"If God (Nature?) had meant man to fly, he would have given us wings." Various earlier reproductive technologies—artificial insemination, in vitro fertilization, surrogate mothers, sperm banks—have elicited similar shock reactions, until we have grown accustomed to them. (Recall the tabloids of the 1950s screaming about "test-tube babies," and excoriating the technology.)

When we do not understand a new technology, we are repulsed by and fearful of it, especially a biological technology, and call it an "ethical issue." But familiarity breeds acceptance in the absence of catastrophic or articulatable moral problems. When we think of cloning humans today, we think of the *Boys from Brazil*, or of identical adults emerging zombie-like and as fully grown adults from pods, with all possessing one mind. What we must do is analytically examine a variety of possible cases of cloning to see if a clear moral problem emerges.

My guess is first of all that the difficulty and inefficiency of cloning, and the correlative expense, will deter people from seeking it, as will societal revulsion (grounded or not). On the other hand, such revulsion will probably diminish after it is revealed that some beloved figure, say a movie star or Mother Teresa, was in fact generated by way of cloning. Finding out that someone was cloned well after the fact, when we already love and admire them, can well defray our revulsion. On the other hand, labeling a child as a "clone" from birth, and marginalizing him or her will almost certainly serve as a self-fulfilling prophecy, making that person "not quite like us" or "freakish." Obviously, such a label would warp a nonclonal child if she were so identified, as well as one derived by cloning.

In discussing cloning of humans with one of my colleagues, Dr. Frank Garry, we sought cases of cloning that were clearly bad, as well as cases where our revulsion was tempered by other characteristics of the case. Clearly a person who cloned himself for spare parts would be viewed as morally bad. An infertile couple who cloned the husband or wife would be viewed as less bad, though probably subject to questions like, "Why not adopt?" a question they could possibly answer in terms of extending their genetic longevity and influence on history. We would welcome the cloning of Mother Teresa more than that of Hitler, especially when we realize that Mother Teresa's special characteristics will surely be otherwise lost (though there is no guarantee that they will be preserved by cloning). And few people without a theological axe to grind would not at least be equivocal about the following case:

Imagine a couple who have struggled to have a child naturally. The child is born, and they can never have another. The child, at one month of age, is struck by lightning in a freak accident, and is dying and irreversibly comatose. The child has not developed a personality yet, and they wish to replace him. Would they be justified in cloning another child from the dying child's tissue? (It's not as if the new child would always be measured against the idealized and deified dead child—that would create moral problems in terms of unfair demands on the new child.) Basically, they want to have a child of their own. They are not cloning themselves, only the dying baby. Intuitively, few of us would say they are *morally wrong* to create the clone, though we might still feel some squeamishness. If the technology is used judiciously and sensitively, I believe that the squeamishness will eventually vanish, as such children grow up normal and healthy.

U

Let us take stock of our discussion. There seems to be nothing inherently wrong with cloning either animals or humans, though clearly the uses to which the technique is put may raise moral issues. In the case of animals and humans, we must worry about the unexpected harmful effects of cloning on an organism, which effects could conceivably manifest themselves later in life. In the case of animals, more so than humans, we must worry about genetic diversity and monoculture. In the human case, an occasional clone here and there should not harm the gene pool any more than twins do. In fact, we probably harm the gene pool far more when we treat the symptoms of genetic diseases and allow people to pass them on. The issue of spare parts is easily handled by our current social ethic. There are probably psychological effects on both parents and children of creating offspring by cloning, but that is also probably true for other reproductive technologies, especially, one would think, the use of donor sperm. The cloning of animals is indeed part of their commodification, or part of the mind-set that created commodification, but clones need not be treated as commodities any more than farm animals need to be.

In sum, in my view, the most important thing about the cloning of Dolly is that it forcefully brings home to us how far behind our technological capability our ethical thinking is. In the absence of time for reflective thinking, we knee-jerk, jump to conclusions, and confuse the aesthetic with the ethical. If science wishes to continue to enjoy public support and preserve its autonomy by being in accord with social ethics, then it must undertake a task it has historically eschewed—educating the general public on scientific and technological innovations, and leading the discussion concerning their ethical import and implications.

NOTE

1. I am grateful to Dick Bowen, Frank Garry, Michael Losonsky, Linda Rollin, Michael Rollin, George Seidel, and the students in my graduate course in Science and Ethics at CSU for dialogue on the issues discussed herein.

REFERENCES

CNN/Time Poll. *Most Americans Say Cloning Is Wrong.* March 1, 1997.

Garry, F., R. Adams, J. P. McCann, and K. G. Odde. "Post-Natal Characteristics of Calves Produced by Nuclear Transfer Cloning." *Theriogenology* 45 (1996): 141–52.

Pursel, V., et al. "Genetic Engineering of Livestock." *Science* 244 (1989): 1281–88.

Rollin, B. E. *The Unheeded Cry: Animal Consciousness, Animal Pain and Science.* Oxford: Oxford University Press, 1989.

———. *Farm Animal Welfare.* Ames, Iowa: Iowa State University Press, 1995.

———. *The Frankenstein Syndrome: Ethical and Social Issues in the Genetic Engineering of Animals.* New York: Cambridge University Press, 1995.

HUMAN CLONING: THE QUESTIONS OF DIGNITY AND IDENTITY

III

We come now to the first of the sections directly concerned with human cloning. Probably the biggest worry that people have is Aldous Huxley's worry, namely that through cloning people will lose their identity. There will be a trivializing of dignity. Humans will become indistinguishable, mass-produced products. This is the concern of the articles in this section. Science-fiction writer Douglas Coupland offers us an amusing but thoughtful picture of a world of cloning, much in the tradition of Huxley, showing just how awful such a practice might become if there are no social or legal sanctions or regulations. Perhaps the biggest identity problem is not that through multiplication people will lose their sense of self, but that through multiplication the rest of us will be subjected to copies and yet more copies of the truly dreadful! Not much dignity in Coupland's world.

Axel Kahn raises worries that are shared by many when thinking about the implications of cloning for issues of human identity. He bases his thinking on one of those philosophies mentioned in our introduction, namely the philosophy of the German philosopher Immanuel Kant. Remember how Kant stressed the significance of human dignity, demanding that we treat humans always as ends in themselves and never as means to other goods or ends? Kahn argues that this philosophy rules out any thoughts of using cloned embryos as resources for already existing humans. But he also thinks that a Kantian position throws grave doubt on the moral wisdom of any form of cloning, even if it were done for an ostensibly good reason like overcoming sterility. Although admitting that there is more to a human than its genes, Kahn fears that the possibilities of abuse, with humans created just for the benefit of others, makes cloning a morally unacceptable practice. Individuality must be protected above all else.

Taking a different stand, popular science writer Stephen Jay Gould goes right to the heart of the issue, saying in few words what most of us cannot say in many. He shows that identical twins, even Siamese twins, can have completely different personalities—for all that, they are more alike than cloned organisms. (For a start, identical twins share the same environment from conception.) "Dolly should inspire our interest and our watchfulness, not our loathing, disgust, or heedless rejection." David Elliott follows up on and extends Gouldian-type arguments, having in his sights those who would claim that our uniqueness is "a precious gift of nature" and that cloning would deprive us of that gift. In a subtle argument, Elliott shows that, although prima facie the moral philosophy of the Kant—who did indeed stress the irreplaceable dignity of each person—would seem to support the case for uniqueness, this is not in fact so. Origins through cloning are quite compatible with a Kantian position. More generally, Elliott finds that arguments based on the problems of uniqueness do not stand as absolute prohibitions against cloning, although at the end of his discussion he warns that this is not necessarily to close all moral discussion on the issue.

8 Clone, Clone on the Range
Douglas Coupland

Back when the first news of successful human cloning was announced, humanity split into two irreconcilable camps: those who said, "How demonic!" and those, like myself—beloved and durable film star Corey Holiday— who said, "Hey! Where do I send my money?" In those glorious late-1990s days of film screenings, animal-rights rallies, and fragrance launches, guests at events invariably divided into the anticloners, with their earnest discussions of ethics, inbreeding, and hillbilly'ed gene pools, and those like myself, so eager and so thrilled to be able to bring humanity the gift of such tried-and-true looks, talent, and industry savvy.

It was a heady era. Overnight it felt as though so many aspects of life were changing: cremation became a thing of the past as franchised DNA storage-facility stocks became the afterworld darlings of NASDAQ; the cost of most medicines fell to the price of a Mars candy bar; and meat became much tastier. Lawyers experienced what can only be described as a renaissance as all dilemmas of law—particularly entertainment, copyright conveyance, deeds, and titles—underwent profound rethinking.

Of course, as the years wore on, the hubbub died. And it was at this time that my poor sweet face, while not becoming fully haggard, was definitely looking somewhat... puffy. Even worse, it was showing on film. The dailies can be cruel.

Makeup calls got earlier and earlier. One box-office flop and—boom!—I'd enter the never-to-return ghetto of geriatric buddy comedies. Yikes.

Yet as time ravaged my looks, I predicted to anyone who might listen that entrepreneurs in retail human cloning would emerge quickly enough. And so they did. First in abandoned Indian Ocean oil rigs and Antarctica—and then slowly and discreetly in more traversed parts of the world.

It was at this point that I, Corey Holiday—magnificent, admired, talented, and feted the world over—after countless years of enthusiastic compliance with the rigors of beauty and the surgeon's scalpel, decided at age fifty it was time to obey Mother Nature's gentle call.

I quietly checked into an exclusive (naturally) cow-based Saskatchewan cloning spa—a spa combining the best of Saskatchewan's cattle country with Canada's lax cloning laws. My public relations staff told folks I was up in the fresh air of Lake Tahoe battling chronic-fatigue syndrome triggered by silicone migration—a plausible alibi if ever there was one.

The spa's rates were steep, but its results were guaranteed. Only superior cattle with modified immune systems were used—cows being the cross-species surrogate of choice. (No cow will ever phone the tabloid press with juicy palimony exclusives.) Clonees were allowed up to five babies per surrogate mom (no womb sharing). Those wishing more than five received generous volume-discount rates.

Myself? I chose five. A single clone might take a dislike to me—and then what? Besides, if I wanted just one kid, why not go out and have one the normal way? The whole point of this procedure was to have lots of exact genetic copies of me to create a flock of worshipful children who would love me as much as I'd enjoy watching them worship me.

Regulations required that we remain at the spa four weeks, lest new tissue samples be called for or some other dreary flaw need mending. The spa itself was bags of fun. Most evenings felt like the Polo Lounge in the old days, and dinner was as star-packed as Morton's on a Monday night.

Thus the snowy Canadian winter passed in a zing. One unexpected treat for me was the arrival, shortly after myself, of veteran film star Lori Breckner, who had been my date for the 1998 Academy Awards ceremony, and who played opposite me in the critically successful box-office dud *Car Crash 500*. ("Yes, Don, I know movies are young young young. But what do a bunch of brats in Glendale know about pain?")

Oh, it was a happy, happy time. Lori and I would sit by the windows, sharing our hopes and dreams about how much our new children would love us, of how we could steer them away from certain types of drugs that they might have too much fun with and toward those cosmetic procedures that would flatter their looks. "Imagine," Lori dreamed aloud one night. "Knowing what seasons your colors are before you're even born! Lucky, lucky children."

While sipping Reverse-Scriptase martinis, Lori and I glanced outside to see the hundreds of beautiful Hereford mommies, glorious and dumb as posts under the great Canadian sky, chewing vitaminized, antibioticized alfalfa while inside each of them our own future little fan clubs incubated. "Look, over there, the one with a white patch on the eye, No. 388—that's yours, honey!" Bliss.

Lori and I discussed how we would transmute all our self-knowledge into our clones so completely that when we died we would technically still be alive—our "death" merely being a technical bookkeeping notation. Imagine feeling as if you are sharing a soul with five others! Lori was indeed a special woman to me. She was the only one I'd met who could connect with me on my own level. We were fated for each other.

And then came that dark morning when we stepped down for coffee and brioches to see the staff aflutter, alarms flaring like hangovers, and a platoon of Mounties interviewing grieving guests. Other patrons were on the pay phones calling their lawyers to alter their wills. "What's gone on?" I asked a passing nurse. Fretful, she told me the news: cattle rustlers.

Dissolve into: the Chicago stockyards. Cut to:... Sorry about the movie jargon. I can't help it. Being a part of the posse was the most real thing that had ever happened to me. Lori too. We looked at each other and said, "It's just like a movie!" I felt so close to this woman.

Lori and I were on scanning duty, fluoroscoping cattle like airport carry-on bags as they galumphed through our stockyard receiving line—a novel preslaughter activity back then, but now compulsory in the United States and Canada. We found two cows, each containing seven embryos—obviously not ours. These cows were then removed to the BMF, the Bovine Midwifing Facility. Only full gestation would reveal the tots' genetic identity. Software mogul? Pop-song diva? Corporation head? Somewhat like waiting for Polaroids to develop over a period of years.

Shortly after finding the rogue septuplets we learned that our "deluxe" Saskatchewan cloning facility had not embedded locator chips in the cows as advertised. That's when we realized our own mommy cows could be practically anywhere. Were they rustled for their meat? Were they taken by terrorists? Kidnappers? Blackmailers? Adoption agencies wanting only pedigree children?

The media got wind of our story, and the Saskatchewan facility was top news

CLONING

for weeks; no doubt the rustlers would be on extra guard now. After Lori spoke with her crystallographer, ChrySanda, in North Hollywood, we roamed northern Montana on an "energy hunch." When we showed up in small-town cafés and feedlots to show photos of cow No. 388 (Lori's brood) and No. 441 (mine), we invariably created a sensation—old good/evil polarity, plus, well, we were and are stars. Citizens were both righteous and helpful, and we always drove away feeling bathed in love of the common man. Sigh.

Some years passed, and then we got a tip. A garbled cell call told us of a private boarding school and ranch near Bozeman, Montana, where "students" were either exceptionally attractive, exceptionally intelligent, exceptionally devious, or all three. So-called school employees signed draconian preagreements barring them from revealing anything. One had escaped, garnered our cell number from a local Webzine ad, and whispered instructions as dogs barked in the background.

We drove along a thin, wooded road and found the entryway into the ranch: laser-guarded, barbed-wired, and accompanied by the anxious *grrrrrr* of concealed attack Dobermans. A good omen—they had something in there worth hiding. A walk around the property's perimeter at first yielded only more of the same. Then we turned a corner and through the trees saw children playing a game of some sort—little houses moved around a board with sticks. The children spotted Lori and me and several of them came over.

"Hello," I said. "I'm film star Corey Holiday."

"And I'm box-office magic Lori Breckner."

The children stared. Then one efficient-looking boy, eight, tops, said, "Excuse me, do you have an appointment? Is somebody expecting you?"

We were agog. His twin (*ha!*) brother asked, "What might this be regarding?"

The younger girl next to him said, "Geoff, was there a memo on this? I don't remember getting the memo."

"Perhaps you should wait. Would you like a cup of coffee or some water?" asked the first boy.

Lori asked the young girl, "What's that game you're playing over there?"

"That? Real Estate. Its fun. I just traded Amy's air rights in exchange for altering my TV network's 9 o'clock slot." A bell rang. "Have to go now," she said. "Facials and colonics. Hope your next pictures gross well." Two of the youngsters slipped us scripts beneath the fence. Bingo. We knew we'd found our rustlers.

Cloning is old news now. We all live with the new reality: blackmailers holding hairbrushes hostage ("Give us your money or we'll make ten of you")... grandmothers reading bedtime stories to 118 baby grandmas... captains of industry rearranging their wills, deeding everything to themselves down the line forever and always. *Plus ça change, plus ça*—wait, that's not really true anymore!

92

And us? Lori and I married shortly after. It was a big ticket—three helicopters more than my previous wedding. But we didn't go back into movies. Instead, we chose to dedicate our lives, possibly forever, to fighting embryo poaching. Us and our ten beautiful children: Cori, Korrie, Corry, Korey, Korrey, Laurie, Lorrie, Laurey, Lorrey, and Lorri.

9   Clone Mammals... Clone Man?
Axel Kahn

T he experiments of I. Wilmut et al. (*Nature* 385[1997]: 810) demonstrate that
sheep embryonic eggs (oocytes) can reprogram the nuclei of differentiated
cells, enabling the cells to develop into any type. The precise conditions under
which this process can occur remain to be elucidated; the factors determining the
success of the technique, and the long-term prospects for animals generated in
this way, still need to be established. But, of course, the main point is that Wilmut
et al. show that it is possible to envisage cloning of adult mammals in a completely
asexual fashion.

The oocyte's only involvement is the role of its cytoplasm in reprogramming
the introduced nucleus and in contributing intracellular organelles—mainly
mitochondria—to the future organism. This work will undoubtedly open up new
perspectives in research in biology and development, for example, in understanding the functional plasticity of the genome and chromatin during develop-

Reprinted by permission from *Nature* 386 (March 13, 1997): 119. Copyright © 1997
Macmillan Magazines Ltd.

ment, and the mechanisms underlying the stability of differentiated states. Another immediate question is to ask whether a species barrier exists. Could an embryo be produced, for example, by implanting the nucleus of a lamb in an enucleated mouse oocyte? Any lambs born in this way would possess a mouse mitochondrial genome.

The implications for humans are staggering. One example is that the technique suggests that a woman suffering from a serious mitochondrial disease might in future be able to produce children free of the disease by having the nucleus of her embryo implanted in a donor oocyte (note that this process is not the same as "cloning").

WOULD CLONING HUMANS BE JUSTIFIED?

But the main question raised by the paper by Wilmut et al. is that of the possibility of human cloning. There is no a priori reason why humans should behave very differently from other mammals where cloning is possible, so the cloning of an adult human could become feasible using the techniques reported.

What medical and scientific "justification" might there be for cloning? Previous debates have identified the preparation of immuno-compatible differentiated cell lines for transplantation, as one potential indication. We could imagine everyone having their own reserve of therapeutic cells that would increase their chance of being cured of various diseases, such as cancer, degenerative disorders, and viral or inflammatory diseases.

But the debate has in the past perhaps paid insufficient attention to the current strong social trend towards a fanatical desire for individuals not simply to have children but to ensure that these children also carry their genes. Achieving such biological descendance was impossible for sterile men until the development of ICSI (intracytoplasmic sperm injection), which allows a single sperm to be injected directly into the oocyte.

But human descendance is not only biological, as it is in all other species, but is also emotional and cultural. The latter is of such importance that methods of inheritance where both parents' genes are not transmitted—such as adoption and insemination with donor sperm—are widely accepted without any major ethical questions being raised.

But today's society is characterized by an increasing demand for biological inheritance, as if this were the only desirable form of inheritance. Regrettably, a person's personality is increasingly perceived as being largely determined by his or her genes. Moreover, in a world where culture is increasingly internationalized and homogenized, people may ask themselves whether they have anything else to transmit to their children apart from their genes. This pressure probably accounts for the wide social acceptance of ICSI, a technique which was widely made available to people at a time when experimental evidence as to its safety was still flimsy. ICSI means that men with abnormal sperm can now procreate.

Going further upstream, researchers have now succeeded in fertilizing a mouse oocyte using a diploid nucleus of a spermatogonium: apparently normal embryonic development occurs, at least in the early stages. But there are also severe forms of sterility—such as dysplasia or severe testicular atrophy—or indeed lesbian couples, where no male germ line exists. Will such couples also demand the right to a biological descendance?

Applying the technique used by Wilmut et al. in sheep directly to humans would yield a clone "of the father" and not a shared descendant of both the father and mother. Nevertheless, for a woman the act of carrying a fetus can be as important as being its biological mother. The extraordinary power of such "maternal reappropriation" of the embryo can be seen from the strong demand for pregnancies in postmenopausal women, and for embryo and oocyte donations to circumvent female sterility. Moreover, if cloning techniques were ever to be used, the mother would be contributing something—her mitochondrial genome. We cannot exclude the possibility that the current direction of public opinion will tend to legitimize the resort to cloning techniques in cases where, for example, the male partner in a couple is unable to produce gametes.

The creation of human clones solely for spare cell lines would, from a philosophical point of view, be in obvious contradiction to an ethical principle expressed by Emmanuel Kant: that of human dignity. This principle demands that an individual—and I would extend this to read human life—should never be thought of only as a means, but always also as an end. Creating human life for the sole purpose of preparing therapeutic material would clearly not be for the dignity of the life created.

Analyzing the use of cloning as a means of combating sterility is much more difficult, as the explicit goal is to create a life with the right to dignity. Moreover, individuals are not determined entirely by their genome, as of course the family, cultural, and social environment have a powerful "humanizing" influence on the construction of a personality. Two human clones born decades apart would be much more different psychologically than identical twins raised in the same family.

THREAT OF HUMAN "CREATORS"

Nonetheless, part of the individuality and dignity of a person probably lies in the uniqueness and unpredictability of his or her development. As a result, the uncertainty of the great lottery of heredity constitutes the principal protection against biological predetermination imposed by third parties, including parents. One blessing of the relationship between parents and children is their inevitable difference, which results in parents loving their children for what they are, rather than trying to make them what they want. Allowing cloning to circumvent sterility would lead to it being tolerated in cases where it was imposed, for example, by authorities. What would the world be like if we accepted that human "creators" could assume the right to generate creatures in their own likeness, beings whose very biological characteristics would be subjugated to an outside will?

The results of Wilmut et al. undoubtedly have much merit. One effect of them is to oblige us to face up to our responsibilities. It is not a technical barrier that will protect us from the perspectives I have mentioned, but a moral one, originating from a reflection of the basis of our dignity. That barrier is certainly the most dignified aspect of human genius.

Individuality
Cloning and the Discomfiting Cases of Siamese Twins

10

Stephen Jay Gould

The aesthetic and ethical foundation of modern Western culture rests firmly on our belief in the distinctiveness of each individual (although many traditions today—and several earlier Western philosophies—do not accept or cherish such a centerpiece). At the climax of Ibsen's great play, *A Doll's House*, when Torvald importunes Nora not to leave him, with a reminder that she is "first and foremost...a wife and a mother," Nora replies: "I believe that first I am an individual."

But the primacy of individual distinction transcends the preferences of any particular culture. The uniqueness of each individual, and the consequent variability among individuals in biological populations of sexually reproducing organisms, provides the sine qua non for evolution by Darwin's mechanism of natural selection. This primary cause of evolutionary change works in a paradoxical manner. Natural selection can create nothing by itself; Darwin's process works by selective elimination and preservation—that is, by imparting higher reproductive success to a subset of individuals fortuitously better adapted to changing local

From *The Sciences* (September/October 1997): 14–16. Reprinted by permission of Stephen Jay Gould.

environments. Natural selection can only work if the individuals of a population are distinctively different, one from the other— thus providing enough "raw material" to fuel the process of natural selection.

In the lifestyle of vertebrates, sex and reproduction are intrinsically intertwined

Rosamond Purcell, Plaster Death Cast of Eng and Chang, *1992*

—and we often err in assuming that they represent the same biological process, with the same evolutionary significance. But sex and reproduction could scarcely fulfill functions that are more different. Reproduction continues a lineage by making more members; sex provides variability among individuals by mixing the genetic products of two parents in each offspring. Asexual reproduction is far more rapid and efficient (think of a splitting bacterium or a budding hydra) than the *Sturm, Drang,* and prolonged gestation of most sexual systems. But asexual reproduction is also an evolutionary dead end because the numerous offspring of a single parent form a clone with no genetic variability among individual members of the clone, unless new and rare mutations arise.

Most people may not have learned this argument about the biological meaning of individuality, but we do know the pattern in a visceral way that defines our intrinsic concept of "normality." That is, we know that each fellow human being is supposed to be distinctly different from all others—and we become, to use the modern vernacular, "freaked" or "weirded out" when we come face-to-face with a human clone.

Eng and Chang Bunker, born in Siam (now Thailand) in 1811, gave the name of their birthplace to the general phenomenon of conjoined, or Siamese twins— the greatest of all challenges to our concept of individuality. Eng and Chang performed for P. T. Barnum, made a pile of money, married a pair of sisters in North Carolina, and fathered about ten children each. (The

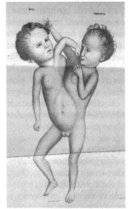

Rosamond Purcell, Ritta and Christina (from Étienne Serres's book, 1833), *1992*

CLONING

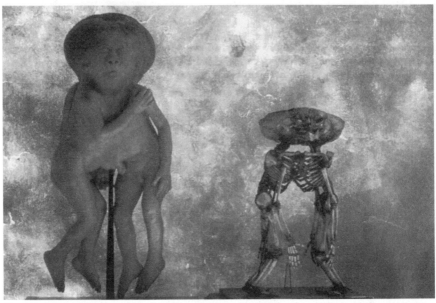

Rosamond Purcell, Wax Model of Twins with Heads Conjoined and Their Skeletons, *1996*

domestic arrangements were, needless to say, unconventional: the wives lived in different houses, while Eng and Chang spent half the week in each domicile.) But Eng and Chang, though a cloned pair of identical twins, were complete and distinct individuals, joined at the abdomen only by a broad band of flesh.

Other styles of Siamese twinning threatened our traditional notion of individuality more directly. Ritta and Christina, born in Sardinia in 1829, had two heads and four arms, but only a single lower body with two legs. When they (or she) died in infancy, probably of overexposure from being exhibited (as the impoverished parents tried to make some money from their misfortune), a Paris newspaper raised the question of the day: "Already it is a matter of grave consideration with the spiritualists, whether they had two souls or one." Most commentators invoked an old Western prejudice and opted for two, on the argument that two complete heads meant two brains and, by implication, two souls. But what, then, could be said of the same phenomenon zipped the other way—that is, a conjoined pair with two lower bodies and only one head?

We regard cloning, or the production of two (or more) genetically identical creatures, as eerie beyond all concept of natural order, at least for mammals and other complex animals. Dolly the sheep, the first mammal cloned from an adult cell of a single parent (and the most famous member of her species since John the Baptist designated Jesus as the Lamb of God), has shocked the world beyond any

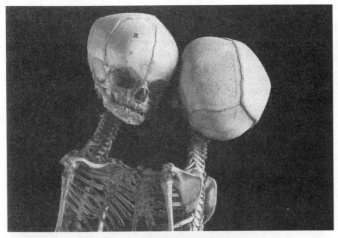

Rosamond Purcell, Twins Joined at the Rib Cage, *1987*

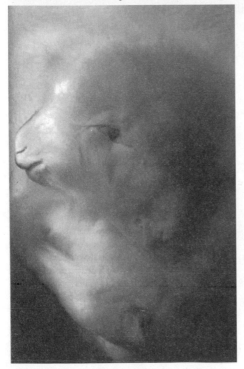

Rosamond Purcell, Two-headed Sheep Embryo, *1991*

merely intellectual reason—primarily by raising for so many people (or so I infer by listening to numerous talk shows and gauging the concerns expressed) our deepest worries about the distinctness of our personhood: Are clones distinct individuals? Does each member of a clone have a soul? Am I still a distinct individual if I clone a daughter from a cell scraped from the inner lining of my cheek?

May I suggest, following the discussion above, that all these fears are misplaced, for these questions have a clear answer, known to all human societies throughout history. Identical twins are clones—in fact, far closer clones than Dolly and her mother (for identical twins share mitochondrial as well as nuclear DNA, and also gestate in the same womb and grow up in the same environment). Yet we know, and have always known, that human identical twins—whatever their quirky similarities in behavioral details as well as physical appearance—become utterly distinct individuals because the unique and contingent pathways of each complex life must impose a separate and formative influence upon any individual, even upon the members of a genetically xeroxed pair. Eng and Chang, closer perforce than any ordinary pair of identical twins, developed distinctly different personalities. One became morose and alcoholic; the other remained benignly cheerful and teetotaling.

Dolly's birth raises legitimate fears—as any powerful new technology must, for all technologies carry the potential for fruitful employment or unspeakable misuse. We can all spell out unacceptable scenarios for human cloning, and we should pursue our ethical debates on this subject with rigor and vigilance. Some nonhuman uses are benign and already in practice. (We have been cloning fruit trees by grafting for decades.) Other potential applications grab my fancy. (I would be powerfully intrigued by a horse race among ten identical Citations—what a test for the skills of jockeys and trainers!)

Dolly should inspire our interest and our watchfulness, not our loathing, disgust, or heedless rejection. Human clones are unambiguous individuals, as proven by the identical twins (or more, up to the Dionne quintuplets and beyond) of all human populations. Environment will impose a cantankerous uniqueness upon any individual. But Darwinian biology—and, therefore, the world of our fathers, our mothers, and all our history—requires the variability conferred by genetic distinctiveness, and therefore defines the domain of all our experience and visceral comfort.

Uniqueness, Individuality, and Human Cloning

David Elliott

The successful cloning of a sheep at the Roslin Institute in Edinburgh has raised the prospect of human cloning[1] to a level that many people—including a U.S. president—find alarming.[2] Even prior to the Roslin announcement in February 1997, Canadian legislators had been working on legislation that would ban almost any attempt to develop human asexual reproduction techniques.[3] Many other European countries also seem to have taken the same approach.[4] Many of these efforts seem motivated by the idea that, even apart from the possible human and social consequences of such technology (popularly believed to be extremely dangerous), there just is something deeply morally objectionable with the entire business of "duplicating" individual human genomes.

My purpose here is to challenge this negative, deontologically motivated assumption. To do so, I want to focus on the most broad, and hence the most minimal, description of what human cloning involves and then try to determine what might be morally objectionable about such an activity. I will maintain that two

Reprinted by permission from *Journal of Applied Philosophy* 15 (1998): 217–30. © Society for Applied Philosophy 1998.

standard counterarguments which purport to show that human cloning is morally objectionable per se are generally unsound. The two arguments are that cloning manufactures (rather than reproduces) persons and that it violates human uniqueness. I conclude that if cloning is morally wrong, it could only be because of present or potential harm imposed on the person cloned, on women who might participate in the procedure or on society generally.

The Definition of "Cloning"

Cloning involves, at the very least, the reproduction of another entity which is, in some sense, identical to an original. Cloning can be achieved, not only with entire organisms, but at the cellular and molecular levels as well. I am, of course, only interested in the first form. The qualification "identical in some sense" acknowledges straightforwardly that it simply would be a mistake to suggest or imply that cloning a human being (or any other animal, for that matter) allows the complete "duplication" or "xeroxing" of that individual. An organism is the expression, not merely of its genome, but of its interactive development within some particular environment. Thus the only thing that is "duplicated" by cloning an organism is that organism's genome, along with those features that are unequivocally tied to its phenotypic expression. In fact, this qualified understanding of what is copied requires even further qualification.[5] Some cloning techniques—e.g., the one used in the Roslin sheep experiment—do not always result in the exact duplication of the cloned individual's genome. The reason for this is that such techniques rely on transferring DNA from a cell nucleus to a donor cell or an (empty) egg. But not all of the operative DNA in an organism is found in a cell's nucleus. The mitochondria are part of a cell that passes its DNA along only in the mother's egg. This mitochondrial DNA apparently does not control very many genes—perhaps only 1 percent of the entire genome—but it does control genes that regulate an organism's metabolism.[6] What this means is that the only ways to get a truly identical clone would be either to have nuclear DNA from a woman put into her own egg or to engage in embryo splitting. (Embryo splitting, which I will explain a bit further below, has the same result because each blastomere that is separated from an early embryo has the same mitochondrial DNA.) In all other cases of nuclear transfer, where another individual's cells or eggs are used, the "copy" is, strictly speaking, genetically nonexact.[7]

There is some resistance in the scientific literature to the straightforward definition that I have just presented.[8] Scientists often tie the definition of "cloning" more carefully to specific techniques that can be used to reproduce organisms asexually that otherwise reproduce sexually. When this is done, the moral issues surrounding cloning seem to change. In view of my general purposes here, however, these more careful understandings do not add up to any important moral differences. Let me briefly explain why.

There are two standard techniques that have been used to clone organisms: nuclear transfer (or transplantation) and embryo splitting (often referred to as blastomere separation). Embryo splitting involves separating cells from very early embryos and then growing them into separate embryos. Because these cells were derived from a single embryo, each separate embryo has the same genome as the original. In 1993 Jerry Hall and other researchers at George Washington University caused great media uproar when they performed this procedure on human cells, but it has been done repeatedly on other animal species since the early 1930s.[9] Nuclear transfer, on the other hand, involves taking a host cell, removing its nucleus, and then replacing it with the nucleus of another donor cell that is to be cloned. Nuclear transfer can be performed using either a cell taken from an embryo (a totipotent cell, i.e., an undifferentiated cell capable of becoming any other cell) or a cell taken from some other part of the body (a nontotipotent cell, i.e., a cell which has differentiated itself into a particular kind of cell, say of skin or muscle). The nuclear transfer of donor DNA from totipotent cells into host embryo cells (blastomeres) has been performed regularly and successfully on domestic animals since the mid-1980s. These techniques were developed from successful cloning experiments done on amphibians in the early 1950s. The breakthrough in the recent Roslin experiment was that apparently for the first time, an adult body cell (a cell from the udder of a pregnant ewe) was used instead of an embryonic cell. Adult body cells are, of course, nontotipotent so the breakthrough here was the demonstration that such cells could be returned to a totipotent state and restart the process of embryonic development.

The point here is that these different techniques could have very different moral implications.[10] First, embryo splitting requires two progenitors; nuclear transfer (in principle) requires only one. Second, embryo splitting produces only limited numbers of the original; nuclear transfer allows large numbers of duplicates. Third, embryo splitting does not entail direct manipulation or selection of genetic material, whereas nuclear transfer does.[11] Since embryo splitting is standardly done on a genetically unique, undeveloped individual, the decision to clone is usually made in the absence of detailed information about what features that individual will come to have. Nuclear transfer, on the other hand, can be done on an adult (as was the case in the Roslin experiment) where selection could be made on the basis of traits that the adult might possess. Hence cloning by nuclear transfer raises more acutely the moral spectre of eugenics, allowing many more opportunities for "positive" or "negative" selection. Finally, embryo splitting is not nearly as technically difficult as nuclear transfer. As a result, it is a much less risky procedure offering fewer chances for things to go wrong. This alone might make a moral difference, since many important concerns about cloning are based on issues surrounding the safety of the procedure. Taking a cell from an adult as opposed to a newly fertilized embryo, for example, means starting a new life from a cell that has acquired genetic mutations over an individual's lifetime. These mutations are often believed to be the basis of both aging and cancer. It is there-fore unclear at the present time whether undue health risks or diminished life expectancy will be one of the serious risks involved in the procedure. Again, embryo splitting, since it proceeds from sexual reproduction, and hence its devel-opment is based on the event of a new and unique genetic code, is much more like the natural procedure of twinning, and hence seems to present us with fewer unknown risks.

The point is that if we were to accept cloning and then pursue the issue of *how* we should clone an individual, the differences between these procedures might become morally important. But since any of these procedures effectively allow the biological duplication of another individual, they raise the same general issue of whether it is right to do this sort of thing. Hence, for my purposes, it is not necessary to attend to the otherwise important moral differences that might hold between cloning techniques.

WHY IS CLONING MORALLY WRONG?

Cloning is an instance of assisted reproduction. As such, it could be morally con-demned for reasons that could be raised against any form of assisted reproduction. I will refer to arguments that employ such a strategy as global anti-assisted repro-duction arguments. These arguments, again, appeal to a *general* position about the immorality of assisted reproduction and then apply this general claim to particular cases of it. Recent philosophical and theological literature on the ethics of assisted reproduction presents many examples of such arguments. Paul Ramsey, for example, a pioneering bioethicist, reasons that since God's creative activity was performed and motivated by love, "...neither should there be among men and women, whose man-womanhood...is in the image of God, any lovemaking set out of the context of responsibility for procreation or any begetting apart from the sphere of human love and responsiveness."[12] The idea here is that since most forms of assisted reproduction are not reasonably seen as expressions of interpersonal love, they become morally suspect. Some feminists, too, seem to defend global arguments against assisted reproductive technologies. A standard argument here is that most forms of assisted reproduction are morally wrong or should be banned because they play into existing patterns of patriarchal power and sexist oppression. The main work of such an argument then becomes showing how this general principle applies to each particular case of assisted reproduction. Whatever the value of this type of argument, I mention it only by way of contrast with the approach that I want to adopt here. My approach begins by looking much more specifically at the particular case of assisted reproduction and then trying to determine whether this would present moral concerns, even a positive social context where sophisticated social or legal regulations might surround the practice, and where there are no serious, known harms associated with it. And, again, what I want to defend now is the claim that human cloning passes this methodological threshold.

If cloning is to be singled out among the forms of assisted reproduction as being wrong per se, its wrong-making property must be connected to the fact that cloning is minimally the attempt to intentionally duplicate, as closely as possible, some individual's genome. Why would this intrinsically be wrong or evil? Two main responses are standardly given to this question: Cloning is wrong (1) because it involves the intention to produce a person with a certain genome, and (2) because it violates some right or value that humans have in being biologically unique. I will refer to the former as the manufacturing argument, and the latter as the uniqueness argument (or the nonuniqueness objection). I will consider each of these proposals in turn.

THE MANUFACTURING OBJECTION

When you think about it, cloning a human would be a very simple, and yet enormously efficient way to select for an individual with certain biological features. The other way to do this would be to engage in genetic alchemy—gene therapy as it is now (perhaps euphemistically) called—and try to change that individual's genome. But this technology is nascent, uncertain, risky, complicated, enormously expensive, and to date, not terribly successful. This fact alone seems to be why certain forms of mammalian cloning in domestic animal husbandry have been largely favored over biotechnology that involves manipulating genomes. You simply find the cow that you like, and then go about producing "copies" of it. Human cloning would seem directly to involve these same "manufacturing" or selection opportunities. Indeed, someone might suggest that these capacities are built right into the decision to clone. It is inherently a decision to produce an individual of a certain type, with certain features that we hope are based in his or her genome. This is how Jeremy Rifkin, a popular biotechnology critic, sees the matter. As he puts it: "It's a horrendous crime to make a Xerox of someone... [because] you're putting a human into a genetic straitjacket. For the first time, we've taken the principles of industrial design—quality control and predictability—and applied them to a human being."[13] The moral idea here seems to be that in manufacturing people, we devalue them; we treat them as objects to be designed rather than as potential subjects or agents capable of their own making.

There are several familiar problems with this familiar line of argument. First, let us assume that the decision to clone is inherently a manufacturing choice (I will question this assumption in a moment). The problem is that these sorts of choices seem to typify so many choices that people make which, even if they are not conscious choices of this sort, at least have the effect of shaping or selecting the traits of their child. The process, for example, of selecting a partner where children could arrive in the future, while it surely is not (and I would hope it never should be) *merely* a decision to select the traits of one's children, it does in part present an opportunity for just this sort of selection. Additionally, if we really do believe that "manufacturing" choices are morally objectionable, then it becomes difficult to imagine why people seeking to adopt a child should have to be consulted about the adoption of any particular child once they have expressed a general interest in adoption. The same might hold for a woman seeking to have a child through artificial insemination by donor. It seems rather strong to hold that her moral qualities as a future parent are seriously diminished by any interest in the general features of the donor, or that she would be an ideal parent if she were to accept sperm only if she could know nothing at all about the physical appearance, family medical history, etc., of its donor.

Furthermore, even if it were just false that people really do make many choices that have the effect of trait selection prior to the birth of their children, they certainly do go to considerable trouble to see that their children develop certain traits after they are born. All of this "manufacturing" seems appropriate, however, given certain standard moral assumptions—e.g., when it is at least not harmful to the child, when it does not severely restrict her opportunities to become an autonomous self-affirming individual or when it is in her interest to have (or avoid) certain characteristics (say, a debilitating disease). The same could be said about preconception selection decisions. Many of these decisions could indeed be frivolous, selfish, and so on. But many of them might be capable of moral defense by showing how they might be important for cultivating the capacities and opportunities for a person's self-development.

What tends to upset many of us about manufacturing decisions, I suspect, is not really that *some* of them might occur in ordinary choices about having a child, but rather that *too many* of them might be present. What might be objectionable, then, is the total quantity of the same sort of choice. Too many manufacturing choices, it might be suggested, push us to the point where we would be treating a (potential) child as an object of his or her parent's desires and goals, rather than a person in his or her own right. Furthermore, being able to determine a person's traits to a very considerable extent might raise concerns about whether that parent is capable of valuing or loving a child unconditionally, or loving him for the person that he might become through self-development. It could also raise questions about whether the potential person would be able adequately to develop her own sense of self and personal agency. In this regard, Joseph Fletcher, another early bioethicist, surely overstated his response to the manufacturing argument when he enthusiastically *celebrates* our potential to manufacture people. "Man," he writes, "is a maker and a selecter [*sic*] and a designer, and the more rationally contrived and deliberate anything is, the more human it is." Fletcher even goes so far as to claim that laboratory reproduction is "radically human compared to conception by ordinary heterosexual intercourse" because the manufacturing is willed, chosen, purposed, controlled; it is a matter of "choice, and not chance."[14]

Whatever merit there might be in responding to the manufacturing objection by arguing either that manufacturing is a standard decision most parents make or that it is not itself morally objectionable, let us for the moment set both of these suggestions aside. There is another consideration that is, I would suggest, even more decisive. The decision to clone is not *inherently* a choice to manufacture a particular individual in a certain way, even though this consequence may be foreseeable. It can simply be a choice to have a child of one's own in the only way possible. The familiar examples standardly offered in the literature as morally defensible reasons to clone illustrate this point. These examples can usually be classified under two main categories: (1) the prevention (by bypassing) of infertility and (2) the avoidance of genetic disease. Given either one or both of these situations, some cloning technique may be the only way that some people might be able to have children of their own. In these cases, however, the decision to have a child could be only that of having a child of one's own; it need not be a detailed manufacturing decision—a point which seems particularly true with respect to bypassing infertility. This is just the sort of outlook that we might arguably suggest is the ideal outlook that couples having children through sexual reproduction should have. Cloning, of course, does come with biological foreknowledge; one would have a fairly good idea of what the child's genome would be, and all that this entails. But, again, simply because there is foreknowledge that a duplication of one's genome will be the outcome of one's decision to have a child, it simply does not follow that the decision to have a child involves or entails a determination to have a child for these reasons. Imagine an analogy with a couple where infertility and genetic disease is not a known consideration, but who may know in advance (say due to some established medical condition) that all of their offspring will be female. It is not obvious that their choice to have a child should be regarded as an instance of sex-selection. If this is right, then there is simply no reason to regard all instances of cloning as instances of manufacturing humans. Other more general, recognizable, and morally defensible motivations, I would suggest, can be present.

THE NONUNIQUENESS OBJECTION

The other main reason that many bioethicists give as a principled objection to cloning is that it creates a biologically nonunique individual. Daniel Callahan, along with other prominent bioethicists, has strenuously voiced this particular concern. He writes:

> For all of its haphazard qualities, there is one enormous advantage in the current lottery: save for the occasional natural twinning, it gives each of us our own unique identity. There is no one else in the world like us. This is a precious gift of nature. It allows us to become our own person, to have some of our parents' genetic traits, but to have even more of our own. Nature does not make us in our parents' image; it makes us in our own unrepeatable image. Cloning would deprive the products of an engineered conception of that gift.[15]

There are two ways of reading Callahan's statement here. He could be claiming either that human cloning would result in (1) an *objective* loss of unique-ness/individuality for the clone or (2) a *subjective* (i.e., a perceived or believed) loss of uniqueness or individuality by the clone. In either case, of course, Callahan seems to be suggesting that the respective loss of uniqueness or individuality is morally significant. The most natural interpretation of Callahan's argument seems to be (1), but as we shall see, this really is an implausible line of argument. I will also argue that even if (1) were true, and human cloning could result in dupli-cating persons and their characters or temperaments, cloning would still not be morally objectionable (at least not solely on the grounds that it creates similar per-sons). I will also hold that while it seems reasonable to think that (2) is empirically unlikely, it seems more plausible to expect that at least some clones would come to feel this way about their situation. I will maintain, however, that this feeling, even if it occurs, is still not a sufficiently morally serious reason not to clone—something more serious would need to be presented, and this probably cannot be found solely in an argument from the moral value of uniqueness.

I have already suggested above why it is a mistake to believe that human cloning would result in an objective loss of uniqueness/individuality for the clone. The reason is that if this claim were true, some very strong kind of biological determinism would also have to be true. For the only thing that we are really considering in cloning humans is the duplication of an individual's genome (and perhaps this too is only partial). So the only way that we could duplicate a person by duplicating that person's genome would be if most, if not all, of what makes up a person comes from the genes. Although there are exceptions, most biologically inclined psychologists seem to resist this idea, and posit that environment plays some role in shaping personality. Furthermore, however similar two persons might be genetically, they clearly would have different experiences, and it seems reasonable to believe that these different experiences would have some differential effect on a person's character and personality.

My focus in this paper, of course, is not on psychology or philosophy of mind. So I do not want to defend rigorously the claims that I have just suggested. Let us, then, for the sake of argument assume that in duplicating a person's genome, we really would be duplicating a person in some interesting sense. Is there something morally objectionable about this? I do not believe so, for three main reasons. First, Callahan's suggestion that nonuniqueness implies repeatability does not lead in any interesting philosophical direction. Secondly, if we accept that nonuniqueness is a moral wrong or evil for persons, then we must entertain rather bizarre beliefs about the moral status or situation of natural clones—i.e., twins, triplets, etc. Underlining both of these arguments will be a third reason—viz., that individuality or uniqueness yields no interesting account of what makes persons or their lives morally valuable.

As we have seen, Callahan suggests that there is a connection between the idea of an individual's uniqueness and the idea of that individual's nonrepeatability. If the possibility of repeating another person is to be morally troublesome, it must imply that in being repeated (or in being rendered nonunique) the original and the repeated individuals are somehow devalued as persons. This suggestion, however, is going to need some further explanation, since it is not obvious why nonrepeatability should be considered a morally significant feature.

One way that it might take on some importance would be to suggest that a person's repeatability implies that he or she is replaceable. Viewing a being as replaceable does not seem to denote a very strong notion of that being's moral worth or value. It seems to imply that we would not be doing anything wrong if

CLONING

we "removed" a particular individual provided that we "replaced" that individual with a new one. Intuitively, at least, many people (I include myself) are unwilling to consider persons in these moral terms, and hence do not want to think of humans as beings whose value as individuals is replaceable. So one way of expanding on Callahan's notion of human value being linked to an individual's nonrepeatability is to claim that individuals have value or status when they are viewed as being nonreplaceable.

This is the way that Immanuel Kant has conceived of the value of persons. In the *Groundwork for the Metaphysics of Morals*, Kant notes a difference between the concepts of the value and dignity of persons. He claims that the idea of value involves the notion of replaceability. Something that is of one value can always replace some other thing of the same value. Value, then, is something that admits of equivalence and hence of exchange. But exchange value is only relative value—i.e., value relative to the interests that some other being takes in that thing. When something possesses dignity, however, it is not merely valuable, it is beyond all value. Such a being is, therefore, not replaceable. It seems appropriate, then, to think of persons as nonreplaceable, and hence to think of them as possessing a dignity rather than intrinsic value.

The argument from uniqueness suggested by Callahan implies that we must think of persons as unique—i.e., that we should find their dignity in their uniqueness. It is worth noting that this is a considerable distance from Kant's general position, since Kant insists that the idea of dignity is formal and abstract. The idea is not that every person is valuable in virtue of being *empirically* different from everyone else. Rather, dignity is grounded in the rational capacity to conform one's will to the moral law, and there is no reason to think that this capacity would be diminished by duplication. So for Kant, dignity certainly comes from being nonreplaceable, but this is simply not the same thing as claiming that every human being finds their dignity in being nonidentical.

Defenders of the argument from uniqueness (like Callahan) seem to extend to Kant's notion of dignity much further than what he had in mind. They maintain that humans are properly seen as possessing dignity rather than merely possessing value, that this dignity comes from being nonreplaceable, and that the only way we can properly understand nonreplaceability is in terms of genetic uniqueness. That is, the argument from uniqueness contends that humans are rightly seen as nonreplaceable only when they are valued as unique, individual, persons. If they were the same, then they would not really be nonreplaceable.

This, however, is not a promising line of argument. It is simply implausible to conceive of nonreplaceability solely or even mainly in terms of uniqueness. This, again, is not the way that Kant thinks of the matter, and since he defends the idea of human worth as linked to the idea of nonreplaceability, his position is worth considering as an alternative. Kant claims that our dignity comes from our capacity for rational agency, for legislating our lives by universal moral laws. In fact, he insists that it is morality—or moral action—which has dignity. Humans acquire dignity by being rational beings who can act morally. As he puts it:

> Nothing can have any worth other than what the law determines. But the legislation itself which determines all worth must for that very reason have dignity, i.e., unconditional and incomparable worth; and the word "respect" alone provides a suitable expression for the esteem which a rational being must have for it. Hence autonomy is the ground of the dignity of human nature and of every rational nature.[16]

The point here (quite apart from Kant's more controversial claim that only the moral law has dignity) is that psychologically normal clones would surely have the capacity for moral agency which Kant refers to here and hence, under his moral schema, would possess dignity. The problem with the uniqueness argument, then, is that it is far too narrow in its implications for an account of the moral status or worth of persons.

What would be helpful now would be a thought experiment imagining the situation that clones might find themselves in, and how we might regard their moral value. The fact is, however, that we really do not need to speculate about such matters. Cloning, even in human reproduction, is a naturally occurring event. Identical twins or triplets regularly appear in the human population. And it is important to stress just what a threat natural clones (monozygotic twins, triplets, etc.) present for concerns about the moral value of individual uniqueness. They are clearly the most serious threat to human uniqueness that we can find, much more so than would be the case with technologically assisted clones. Monozygotic or "identical" twins not only share the same nuclear DNA, but have the same mitochondrial DNA as well.[17] They are usually gestated in the same woman at the same time and circumstances in her body, and are standardly raised in the same cultural and familial contexts. So with natural clones (identical twins) both genetic and environmental forces standardly conspire to make them relatively nonunique individuals.

In contrast, clones resulting from technically assisted reproduction would probably not face this convergence of standardizing forces. They might not, and probably would not, be strictly genetically identical (e.g., they could easily not share the same mitochondrial DNA); they probably would not have the same gestational mother at the same time and circumstances in her body, and they could be raised temporally and spatially apart from their "siblings." The point here is that if cloning is morally objectionable because it would produce nonunique human beings, then it must be the case that identical twins are either (1) less morally valuable or (2) somehow worse off than the rest of us.

Before we consider the defensibility of these two claims, we should pause and reconsider the analogy that I am drawing between technologically assisted clones and natural ones. My claim is that if there is a serious threat to nonuniqueness worthy of our moral attention, it seems to point more to natural rather than technically assisted clones. Even if someone accepts this argument, however, it might still be pointed out that, setting aside differences in the technical methods involved in cloning (for reasons that I have set out above), a significant moral difference between these two cases still remains. Whereas the production of nonunique genomes in technically assisted cloning would be intentional and deliberate, in most cases of natural twinning it is not. Moreover, it seems that in many contexts we do right by accepting circumstances as we find them, but it would often be clearly wrong to set out to bring about such circumstances. Thus the analogy that I want to draw between technically assisted and natural clones (twins) fails because there is a relevant moral difference between these two cases: intentionality.

This objection, however, really does not point to a significant disanalogy. The problem is that intentionality and deliberateness do not, *on their own*, add any moral significance. The only way that they could would be if natural twins either possessed a different moral status or are somehow worse off than genetically unique individuals—i.e., only if either option (1) or (2) above is true. If the two cases are otherwise morally similar, then intentionality points to no important moral difference between them.

To clarify this argument further, consider a similar analogy. Imagine two situations, one of a couple who, through no avoidable fault, awareness, or intention of their own, have a child with a severely disabling genetic disease, and another of a couple who, by clear intentional means (say by direct genetic manipulation of an embryo), set out to have a child with the same disease. We probably would want to say that the former couple has not done anything wrong, but the latter arguably has. But the mere fact that in the latter case the action is intentional adds no moral weight unless it is already true that the disease makes either child somehow worse off than healthy individuals. If all else is equal between two cases like this, it is simply unclear what makes intentionality morally significant. Intentionality is morally significant only if there is an intention to do something that is otherwise morally good or bad, right or wrong. The analogy, therefore, between natural and technically assisted clones stands up.

We are left, then, with suggestions (1) and (2). If twins are somehow morally worse off than the rest of us, we need to be shown exactly what it is that twins are deprived of and its moral nature needs to be explained. It is very difficult to accept that twins are deprived of something essentially related to their inherent moral value as persons. And it does not seem any more obvious that in not being biologically unique twins are made worse off.

It might be suggested that they are psychologically damaged in some way because of their biological nonuniqueness. But evaluations of the "similarity" of twins are often psychologically and morally ambivalent. In many cases this similarity can be an *advantage* in that they can share an enviable sense of empathy and human relatedness. Given this ambivalence, it is not clear how one could strongly maintain that nonuniqueness is objectionable, something to be avoided as a matter of moral principle.

It could be simply denied, of course, that twins or triplets are morally worse off than the rest of us. But if the defender of the argument from uniqueness were to do this, she would have to claim something like the following: Twins, although biologically identical, have not been wronged or are not worse off than people who are unique *and* being nonunique is morally worse than being unique. This is, of course, inconsistent. To address this inconsistency she might claim that what is wrong here has nothing to do with the twins, their state of mind, or even their state of being, but (again) with someone else's *intention* to create biologically nonunique individuals. But now it is just not clear what work the appeal to nonuniqueness is doing in the overall argument. If there is nothing wrong with (the fact of) nonuniqueness itself, it is not clear why intending to do it would be wrong. That is, it is not obvious how some action or state of affairs, X, could *never* be bad (either instrumentally or intrinsically), whereas intending to do X or to bring about X, would *always* be wrong or bad. As a result of this problem, I see no sound way for someone to defend both (a) the claim that twins are not worse off than the rest of us in being nonunique and (b) the claim that nonuniqueness for clones is morally bad.

Cloning may diminish the cloned person's subjective sense of uniqueness, but no more, perhaps, than an adopted child's sense of alienation when she is totally genetically unrelated to her (social) parents. The threat of this sense of alienation and lack of self identity—and I admit it can often be very real and painful—do not seem to be any more sufficient reasons for arguing against adoption than is the risk of a lack of sense of uniqueness which might arise in a person who is totally unrelated (genetically) to her mother.

Adoption, of course, is not strictly analogous to the decision to clone. Adoption is the best solution to an apparently much worse situation—i.e., being raised without a deep, lasting relationship with at least one adult guardian. As such, the risk of self-alienation may be worth taking because the alternative is much worse.

CLONING

Cloning, on the other hand, is not the solution to a graver outcome. The alternatives are either not being born or parents not having genetically related children—neither of which seems nearly as bad as a child going through her childhood without a relationship to a loving and committed guardian (or guardians). But this difference does not seem important. At least it does not take away from my claim that the risk of feeling nonunique is no more morally decisive against cloning than is the risk of feeling alienated in adoption.

It is also worth stressing that, as the practice of adoption shows, the relation of father or mother and child is also a profoundly *social* relation; it cannot be reduced merely to biological relations. There seems to be no compulsion for a child who is a clone of the person who has raised her to understand or interpret that relation as a sibling one. And it is not clear that the cloned individual, like any other natural twin, needs to see her individuality as importantly connected to her genome, or even to phenotypic similarities with others. Furthermore, there is some evidence that identical twins separated from birth tend to show comparatively more divergent personalities and behaviors than twins who are not separated.[18] It seems reasonable, then, to believe that identical twins separated by temporal distance will show even more differences. Hence a psychological sense of uniqueness may not be seriously threatened. All of the differences implied in temporal distance would probably give rise to sufficient grounds for a feeling of difference. Thus it seems that the claim about felt identity being psychologically damaging is not very plausible.

The argument that I have just made, however, is an empirically based one, and I could simply be wrong about the relevant facts. Even if this kind of negative feeling with regard to one's individuality could arise, it still does not seem that it *must* arise, nor that if it does arise, it raises serious moral concerns in relation to the basic value of persons. This lack of necessity is morally relevant because in the case of most adults it is the individual him/herself who is probably best seen as the person who takes some final sense of responsibility for self-understanding and self-development. Where this is not an appropriate allocation of responsibility, this is usually due to the way that the individuals have been raised by their parents or guardians. Parents can, and often do, raise genetically nonidentical children to be and appear like other persons. There is surely no need to clone someone in order to threaten his feeling of uniqueness and individuality. A parent, then, who is acting as I would suggest a good parent should, will address this issue, and will encourage difference and individuality.

Suppose, however, that a cloned child has all of these environmental counterbalances in place and still comes to feel nonunique. It is again not clear that this consequence finally renders cloning immoral. The problem now is that whatever the foundation of an individual's inherent moral worth is, it surely must not rest importantly on being genetically different from other people. If it did, then the disparate value of twins problem that I have just been discussing resurfaces. Since there seems to be every reason, intuitively at least, to regard twins as possessing equal moral status or value as the rest of us, we should reject the idea that there is anything of any great moral import to be found in possessing different or unique genomes (or perhaps even unique phenotypes). So if I am right about the ground of inherent moral worth, duplicating a person's genome (or phenotype) cannot morally violate or affront that worth.

We have seen that even a philosopher with Kantian convictions, who values the role of moral principle to a high degree, might not object to cloning, since he or she would find inherent moral worth in the possession of rational agency. Surely clones would have this in spite of their biological identity. Someone defending the uniqueness objection, however, could put forward a non-Kantian notion of rationality, something much less formal or abstract. Perhaps, to connect back to the manufacturing objection, the claim could be made that a nonunique genome imposes limitations on an individual's capacity for self-development, leading to a sense of disempowerment, and personal inadequacy. It would be odd, however, to suggest that inherent human worth is somehow connected to being born without these limitations. Whether I have this sense of empowerment or not, it seems that it should never diminish my claim for others to respect my value as a person. Forms of social oppression usually have the effect of diminishing self-respect or self-empowerment in their victims which seriously impedes their own self- and social development. But, again, this fact should never detract from their abstract moral worth.[19] Thus it would provide no good reasons for treating a person lacking this self-empowerment badly, nor for holding that such persons should not be born simply because they will come to lack this quality.

CONCLUSION

It is understandable why individual uniqueness might be seen as morally significant. It certainly seems *biologically* important. Evolution by natural selection benefits considerably from individual genetic diversity, and sexual reproduction by and large ensures that this diversity will continue. Hence, given this biological norm, and the role that it has played in our own evolutionary history, it seems understandable why we might see moral significance in it. This may equally explain why many people are disturbed by the prospect of human clones, and for

CLONING

that matter, why we are in awe of twins, triplets, and so on. But claiming that individual uniqueness is of paramount moral importance is another matter entirely, and we have seen that it does not present much philosophical promise as a way of soundly articulating moral concerns about human cloning.

The argument that I have defended here should never provide a defense for further research and experimentation into the prospects of cloning. It is important to stress this point since it might seem like a very easy step from the argument that some particular activity is not wrong in itself to the argument that it is at least not impermissible to try to realize it. This is a misguided assumption, however, for two main reasons. First, it should be obvious that merely showing that something is not intrinsically wrong goes nowhere to show that it would not have morally bad consequences, and if it were readily available, that it would not provide some people with what Ruth Macklin has described as a tremendous "opportunity for mischief."

The second reason is that the research or clinical procedures necessary to provide any opportunities to clone human genomes could simply be immoral. This seems to be the case with human cloning at the present time—especially if the procedure under consideration is nuclear transfer using somatic cell nuclei. The Roslin experiment was so dreadfully inefficient that an attempt to try the same procedure on humans would almost certainly raise very serious moral questions. Ian Wilmut and his colleagues began with 434 eggs—many times the number of eggs in a ewe's lifetime ovulation. Fusion with the DNA from somatic cells was successful only 277 times. Worse still, only 29 of these transfers divided sufficiently for implantation. And of these 29, only 13 were implanted in ewes. Only *one* of these ewes became pregnant and then gave birth to the only successful clone.

Now since domestic animals are much more fertile than humans—perhaps three or four times—it would follow that about 1,200 to 1,500 human eggs would be needed to conduct a similar experiment in humans—several times more than a woman's entire lifetime supply. Furthermore, the procedure would probably require that about fifty women should be willing to be impregnated and possibly give birth at a very high rate of implantation failure and miscarriage.[20] These estimates assume, of course, that there would be no species-specific obstacles in using the Roslin procedure in humans. But there have been such problems. In 1981, Karl Illmensee and Peter Hoppe cloned mice by nuclear transfer.[21] For reasons that are not really understood, other scientists have had considerable difficulties reproducing this experiment. It is widely believed, however, that there is something specific in the reproductive cycle in mice that has prevented the widespread cloning of mice by nuclear transfer. Similar problems might arise if the procedure were tried in humans, resulting in a much more severe rate of pregnancy failure, and of increased health risks for either the mothers involved or the individuals who were cloned.

My point is that these, and many other issues, pose questions about risk assessments which, it seems to me, would never be outweighed by the argument that I have presented above. Thus it would simply be a misuse of my argument to suggest that it in any way implies that research on humans would be justified. Rather, the argument assumes a counterfactual context in which there are no serious risks attending to the procedure of nuclear transfer itself. The counterfactuality of this procedure, however, may not always be present. Sadly, in my view, scientific research proceeds without anywhere near the same moral reservations in its use of nonhuman animals as with human experimentation. It could well be that the gap between the applicability of this procedure in animals and humans could narrow severely through research conducted on nonhuman animals, even primates. There is currently a considerable interest in promoting this sort of research in nonhuman animals in order to produce genetically engineered pharmaceuticals and even organs for human benefit. Indeed, this, and the commercial benefits likely to follow from such discoveries, are the main reasons why the Roslin Institute conducted its experiment in the first place. If so, the possibility of developing this sort of technique through experiments in humans may not present as many difficulties as it now seems to. If I am right about the argument presented above, however, only harm or other consequentialist considerations should give us reasons for moral pause on this issue.

ACKNOWLEDGMENTS

I am grateful to Phil Gosselin, Stephen Clark, and one of the reviewers for this journal for helpful comments on earlier versions of this paper. I am particularly indebted to Eldon Soifer for detailed written comments and lengthy discussions about the issues raised in this paper.

NOTES

1. See Ian Wilmut et al., "Viable Offspring Derived from Fetal and Adult Mammalian Cells," *Nature* 385 (February 27, 1997): 810–13; and K. H. S. Campbell et al., "Sheep Cloned by Nuclear Transfer from a Cultured Cell Line," *Nature* 380 (March 7, 1996): 64–66. I will use the term "cloning" in this paper to refer exclusively to human cloning.

CLONING

2. See Meredith Wadman, "White House Bill Would Ban Human Cloning, *Nature* 387 (June 12, 1997): 644.

3. Bill C-47 "An act respecting human reproductive technologies and commercial transactions relating to human reproduction." The bill, which did not pass, proposed to make it illegal, on a maximal fine of $500,000 or ten years (or both), to "manipulate an ovum, zygote, or embryo for the purpose of producing a zygote or embryo that contains the same genetic information as a living or deceased human being or a zygote or embryo or fetus, or implant in a woman a zygote or embryo so produced" (section 4.1a).

4. See DeClan Butler, "European Ethics Advisers Back Cloning Ban," *Nature* 387 (June 5, 1997): 536; and Robin Herman, "European Bioethics Panel Denounces Human Cloning," *Washington Post*, June 10, 1997, p. Z19. An interesting exception to this is Australia, where the Infertility (Medical Procedures) Act (Victoria) passed restrictive legislation much earlier (1984) than many other countries. This Australian legislation, it seems, was the product of a similar flurry of concern about technological developments in the late 1970s surrounding the success of in vitro fertilization techniques. For more about this legislation see Margaret Brumby and Pascal Kasimba, "When Is Cloning Lawful?" *Journal of In Vitro Fertilization and Embryo Transfer* 4 (1970): 198–204. See also Peter Singer and Deane Wells, *Making Babies: The New Science and Ethics of Conception* (New York: Scribner, 1985), pp. 146–49 for discussion of an important background report on which this legislation was based. In 1990, Britain introduced legislation (the Human Fertilization and Embryology Act) which ostensibly banned human cloning. But there have been concerns raised about whether this law forbids cloning only human embryos, and if so, whether it would thereby allow cloning adults by nuclear transfer as per the Roslin experiment, since such a procedure does not initially involve manipulating embryos. For more on this see Ehsan Masood, "Cloning Technique 'Reveals Legal Loophole,' " *Nature* 385 (February 27, 1997): 757.

5. I will use terms like "duplicate" and "copy" throughout this paper in terms of the two qualifications that I have just noted in the previous sentence.

6. See *A Dictionary of Biology*, 3d ed. (New York: Oxford University Press, 1996), p. 325.

7. For more about such matters, see Richard Lewontin, "The Confusion over Cloning," *New York Review of Books* (October 23, 1997).

8. For more about this dispute see Rebecca Voelker, "A Clone by Any Other Name Is Still an Ethical Concern," *Journal of the American Medical Association* 271 (1994): 331; and J. Cohen and Giles Tompkin, "The Science, Fiction, and Reality of Embryo Cloning," *Kennedy Institute of Ethics Journal* 4 (1994): 194.

9. See J. L. Hall et al., "Experimental Cloning of Human Polypoid Embryos Using an Artificial Zona Pellucida," American Fertility Society conjointly with the Canadian Fertility and Andrology Society, Program Supplement, Abstracts of the Scientific Oral and Poster Sessions, Abstract 0-001, SI (1993).

10. For more on this see National Advisory Board on Ethics in Reproduction (NABER), "Report on Human Cloning through Embryo Splitting: An Amber Light," *Kennedy Institute of Ethics Journal* 4 (1994): 252.

11. An exception to this might be where several blastomeres are produced by separation and then cryopreserved for possible implantation after one of them has developed to a stage where distinct features can be observed and expected. But even here the initial decision to clone by blastomere separation must still be done in the absence of detailed phenotypic information. And this decision would have to be aligned with the decision whether to discontinue cryopreservation or to implant the embryo.

12. *Fabricated Man: The Ethics of Genetic Control* (New Haven: Yale University Press, 1970), p. 88.

13. Quoted in Jeffrey Kluger, "Will We Follow the Sheep?" *Time* (March 10, 1997): 40.

14. "Ethical Aspects of Genetic Control," *New England Journal of Medicine* 285 (1971): 780–81.

15. "Perspective on Cloning: A Threat to Individual Uniqueness; An Attempt to Aid Childless Couples by Engineered Conceptions Could Transform the Idea of Human Identity," *Los Angeles Times*, November 12, 1993, p. B7.

16. *Groundwork for the Metaphysics of Morals*, trans. James W. Ellington (Indianapolis: Hackett, 1993), p. 41 (436, Academy pagination).

17. This point is stressed by Stephen Jay Gould, "Individuality: Cloning and the Discomfiting Case of Siamese Twins, *The Sciences* 37 (September–October, 1997): 16.

18. For more about this see Richard C. Lewontin, *Human Diversity* (New York: Scientific American Library, 1982).

19. I do not intend to suggest here that moral worth should be seen as an exclusively abstract, objective property of persons, and never a subjectively felt or valued empowerment. I only intend to commit myself to the view that an account of moral worth or value should be at least in part objective and abstract.

20. The numbers in this, and the previous, sentence are from Stephen Strauss, "Hello Dolly, It's So Good to See You," *Globe and Mail*, March 1, 1997, p. A5.

21. See Jean L. Marx, "Three Mice 'Cloned' in Switzerland," *Science* 211 (1981): 375–76.

HUMAN CLONING: THE CASE AGAINST

IU

W e come now to the heart of the ethical debate. In a major article, Leon R. Kass argues fervently that cloning humans would be wrong in the very deepest sense. He believes passionately that there is a deep sense of repugnance that such a practice brings forth in a morally sensitive person. This is "the emotional expression of deep wisdom, beyond reason's power fully to articulate it." Some practices—Kass lists father-daughter incest, sex with animals, mutilating a corpse, eating human flesh, or even rape or murder—are simply beyond the moral pale. Cloning humans falls into this category. If we are to have any right at all to respect as responsible moral individuals, we must listen to and be guided by these most fundamental of human emotions or feelings.

Not that reasoned arguments cannot also be given. Most particularly, cloning goes against everything that we know is good and natural and decent about family relationships, both of the kind which extends to children and other relatives as well as that centering on the deepest and most proper connections between men and women. It is a "blatant violation of the inner meaning of parent-child relations." To come together and to have children is to put oneself at the mercy of forces beyond oneself—to accept the risk of whatever might transpire. It is to realize that we are finite beings, not God, and it is to acknowledge that there are limits to what we can control. For these reasons, backed by the strongest of emotions, Kass argues that without exception we must ban all human cloning, we must put away seductive but false prospects of indefinite progress toward perfection—of humans or of anything else—and we must prize and enhance prospects of human dignity. Quoting the late Paul Ramsey, one of the founders of modern bioethics, Kass concludes: "The good things that men do can be made complete only by the things they refuse to do."

12 The Wisdom of Repugnance

Leon R. Kass

Our habit of delighting in news of scientific and technological break-throughs has been sorely challenged by the birth announcement of a sheep named Dolly. Though Dolly shares with previous sheep the "softest clothing, woolly, bright," William Blake's question, "Little Lamb, who made thee?" has for her a radically different answer: Dolly was, quite literally, made. She is the work not of nature or nature's God but of man, an Englishman, Ian Wilmut, and his fellow scientists. What's more, Dolly came into being not only asexually—ironically, just like "He [who] calls Himself a Lamb"—but also as the genetically identical copy (and the perfect incarnation of the form or blueprint) of a mature ewe, of whom she is a clone. This long-awaited yet not quite expected success in cloning a mammal raised immediately the prospect—and the specter—of cloning human beings: "I a child and Thou a lamb," despite our differences, have always been equal candidates for creative making, only now, by means of cloning, we may both spring from the hand of man playing at being God.

Reprinted From *The New Republic* (June 2, 1997): 17–26, with permission of the author.

CLONING

After an initial flurry of expert comment and public consternation, with opinion polls showing overwhelming opposition to cloning human beings, President Clinton ordered a ban on all federal support for human cloning research (even though none was being supported) and charged the National Bioethics Advisory Commission to report in ninety days on the ethics of human cloning research. The commission (an eighteen-member panel, evenly balanced between scientists and nonscientists, appointed by the president and reporting to the National Science and Technology Council) invited testimony from scientists, religious thinkers, and bioethicists, as well as from the general public. It is now deliberating about what it should recommend, both as a matter of ethics and as a matter of public policy.

Congress is awaiting the commission's report, and is poised to act. Bills to prohibit the use of federal funds for human cloning research have been introduced in the House of Representatives and the Senate; and another bill, in the House, would make it illegal "for any person to use a human somatic cell for the process of producing a human clone." A fateful decision is at hand. To clone or not to clone a human being is no longer an academic question.

Taking Cloning Seriously, Then and Now

Cloning first came to public attention roughly thirty years ago, following the successful asexual production, in England, of a clutch of tadpole clones by the technique of nuclear transplantation. The individual largely responsible for bringing the prospect and promise of human cloning to public notice was Joshua Lederberg, a Nobel Laureate geneticist and a man of large vision. In 1966, Lederberg wrote a remarkable article in the *American Naturalist* detailing the eugenic advantages of human cloning and other forms of genetic engineering, and the following year he devoted a column in the *Washington Post*, where he wrote regularly on science and society, to the prospect of human cloning. He suggested that cloning could help us overcome the unpredictable variety that still rules human reproduction, and allow us to benefit from perpetuating superior genetic endowments. These writings sparked a small public debate in which I became a participant. At the time a young researcher in molecular biology at the National Institutes of Health (NIH), I wrote a reply to the *Post*, arguing against Lederberg's amoral treatment of this morally weighty subject and insisting on the urgency of confronting a series of questions and objections, culminating in the suggestion that "the programmed reproduction of man will, in fact, dehumanize him."

Much has happened in the intervening years. It has become harder, not easier, to discern the true meaning of human cloning. We have in some sense been softened up to the idea—through movies, cartoons, jokes, and intermittent commentary in the mass media, some serious, most lighthearted. We have become accustomed to new practices in human reproduction: not just in vitro fertilization, but also embryo manipulation, embryo donation, and surrogate pregnancy. Animal biotechnology has yielded transgenic animals and a burgeoning science of genetic engineering, easily and soon to be transferable to humans.

Even more important, changes in the broader culture make it now vastly more difficult to express a common and respectful understanding of sexuality, procreation, nascent life, family, and the meaning of motherhood, fatherhood, and the links between the generations. Twenty-five years ago, abortion was still largely illegal and thought to be immoral, the sexual revolution (made possible by the extramarital use of the pill) was still in its infancy, and few had yet heard about the reproductive rights of single women, homosexual men, and lesbians. (Never mind shameless memoirs about one's own incest!) Then one could argue, without embarrassment, that the new technologies of human reproduction—babies without sex—and their confounding of normal kin relations—who's the mother: the egg donor, the surrogate who carries and delivers, or the one who rears?—would "undermine the justification and support that biological parenthood gives to the monogamous marriage." Today, defenders of stable, monogamous marriage risk charges of giving offense to those adults who are living in "new family forms" or to those children who, even without the benefit of assisted reproduction, have acquired either three or four parents or one or none at all. Today, one must even apologize for voicing opinions that twenty-five years ago were nearly universally regarded as the core of our culture's wisdom on these matters. In a world whose once-given natural boundaries are blurred by technological change and whose moral boundaries are seemingly up for grabs, it is much more difficult to make persuasive the still compelling case against cloning human beings. As Raskolnikov put it, "man gets used to everything—the beast!"

Indeed, perhaps the most depressing feature of the discussions that immediately followed the news about Dolly was their ironical tone, their genial cynicism, their moral fatigue: "An Udder Way of Making Lambs" (*Nature*), "Who Will Cash in on Breakthrough in Cloning?" (*Wall Street Journal*), "Is Cloning Baaaaaaaad?" (*Chicago Tribune*). Gone from the scene are the wise and courageous voices of Theodosius Dobzhansky (genetics), Hans Jonas (philosophy), and Paul Ramsey (theology) who, only twenty-five years ago, all made powerful moral arguments against ever cloning a human being. We are now too sophisticated for such argumentation; we wouldn't be caught in public with a strong moral stance, never mind an absolutist one. We are all, or almost all, postmodernists now.

CLONING

Cloning turns out to be the perfect embodiment of the ruling opinions of our new age. Thanks to the sexual revolution, we are able to deny in practice, and increasingly in thought, the inherent procreative teleology of sexuality itself. But, if sex has no intrinsic connection to generating babies, babies need have no necessary connection to sex. Thanks to feminism and the gay rights movement, we are increasingly encouraged to treat the natural heterosexual difference and its preeminence as a matter of "cultural construction." But if male and female are not normatively complementary and generatively significant, babies need not come from male and female complementarity. Thanks to the prominence and the acceptability of divorce and out-of-wedlock births, stable, monogamous marriage as the ideal home for procreation is no longer the agreed-upon cultural norm. For this new dispensation, the clone is the ideal emblem: the ultimate "single-parent child."

Thanks to our belief that all children should be *wanted* children (the more high-minded principle we use to justify contraception and abortion), sooner or later only those children who fulfill our wants will be fully acceptable. Through cloning, we can work our wants and wills on the very identity of our children, exercising control as never before. Thanks to modern notions of individualism and the rate of cultural change, we see ourselves not as linked to ancestors and defined by traditions, but as projects for our own self-creation, not only as self-made men but also man-made selves; and self-cloning is simply an extension of such rootless and narcissistic self-re-creation.

Unwilling to acknowledge our debt to the past and unwilling to embrace the uncertainties and the limitations of the future, we have a false relation to both: cloning personifies our desire fully to control the future, while being subject to no controls ourselves. Enchanted and enslaved by the glamour of technology, we have lost our awe and wonder before the deep mysteries of nature and of life. We cheerfully take our own beginnings in our hands and, like the last man, we blink.

Part of the blame for our complacency lies, sadly, with the field of bioethics itself, and its claim to expertise in these moral matters. Bioethics was founded by people who understood that the new biology touched and threatened the deepest matters of our humanity: bodily integrity, identity and individuality, lineage and kinship, freedom and self-command, eros and aspiration, and the relations and strivings of body and soul. With its capture by analytic philosophy, however, and its inevitable routinization and professionalization, the field has by and large come to content itself with analyzing moral arguments, reacting to new technological developments, and taking on emerging issues of public policy, all performed with a naive faith that the evils we fear can all be avoided by compassion, regulation, and a respect for autonomy. Bioethics has made some major contributions in the protection of human subjects and in other areas where personal freedom is threatened; but its practitioners, with few exceptions, have turned the big human questions into pretty thin gruel.

One reason for this is that the piecemeal formation of public policy tends to grind down large questions of morals into small questions of procedure. Many of the country's leading bioethicists have served on national commissions or state task forces and advisory boards, where, understandably, they have found utilitarianism to be the only ethical vocabulary acceptable to all participants in discussing issues of law, regulation, and public policy. As many of these commissions have been either officially under the aegis of NIH or the Health and Human Services Department, or otherwise dominated by powerful voices for scientific progress, the ethicists have for the most part been content, after some "values clarification" and wringing of hands, to pronounce their blessings upon the inevitable. Indeed, it is the bioethicists, not the scientists, who are now the most articulate defenders of human cloning: the two witnesses testifying before the National Bioethics Advisory Commission in favor of cloning human beings were bioethicists, eager to rebut what they regard as the irrational concerns of those of us in opposition. One wonders whether this commission, constituted like the previous commissions, can tear itself sufficiently free from the accommodationist pattern of rubber-stamping all technical innovation, in the mistaken belief that all other goods must bow down before the gods of better health and scientific advance.

If it is to do so, the commission must first persuade itself, as we all should persuade ourselves, not to be complacent about what is at issue here. Human cloning, though it is in some respects continuous with previous reproductive technologies, also represents something radically new, in itself and in its easily foreseeable consequences. The stakes are very high indeed. I exaggerate, but in the direction of the truth, when I insist that we are faced with having to decide nothing less than whether human procreation is going to remain human, whether children are going to be made rather than begotten, whether it is a good thing, humanly speaking, to say yes in principle to the road which leads (at best) to the dehumanized rationality of *Brave New World*. This is not business as usual, to be fretted about for a while but finally to be given our seal of approval. We must rise to the occasion and make our judgments as if the future of our humanity hangs in the balance. For so it does.

THE STATE OF THE ART

If we should not underestimate the significance of human cloning, neither should we exaggerate its imminence or misunderstand just what is involved. The procedure

CLONING

is conceptually simple. The nucleus of a mature but unfertilized egg is removed and replaced with a nucleus obtained from a specialized cell of an adult (or fetal) organism (in Dolly's case, the donor nucleus came from mammary gland epithelium). Since almost all the hereditary material of a cell is contained within its nucleus, the renucleated egg and the individual into which this egg develops are genetically identical to the organism that was the source of the transferred nucleus. An unlimited number of genetically identical individuals—clones—could be produced by nuclear transfer. In principle, any person, male or female, newborn or adult, could be cloned, and in any quantity. With laboratory cultivation and storage of tissues, cells outliving their sources make it possible even to clone the dead.

The technical stumbling block, overcome by Wilmut and his colleagues, was to find a means of reprogramming the state of the DNA in the donor cells, reversing its differentiated expression and restoring its full totipotency, so that it could again direct the entire process of producing a mature organism. Now that this problem has been solved, we should expect a rush to develop cloning for other animals, especially livestock, in order to propagate in perpetuity the champion meat or milk producers. Though exactly how soon someone will succeed in cloning a human being is anybody's guess, Wilmut's technique, almost certainly applicable to humans, makes *attempting* the feat an imminent possibility.

Yet some cautions are in order and some possible misconceptions need correcting. For a start, cloning is not Xeroxing. As has been reassuringly reiterated, the clone of Mel Gibson, though his genetic double, would enter the world hairless, toothless, and peeing in his diapers, just like any other human infant. Moreover, the success rate, at least at first, will probably not be very high: the British transferred 277 adult nuclei into enucleated sheep eggs, and implanted twenty-nine clonal embryos, but they achieved the birth of only one live lamb clone. For this reason, among others, it is unlikely that, at least for now, the practice would be very popular, and there is no immediate worry of mass-scale production of multicopies. The need of repeated surgery to obtain eggs and, more crucially, of numerous borrowed wombs for implantation will surely limit use, as will the expense; besides, almost everyone who is able will doubtless prefer nature's sexier way of conceiving.

Still, for the tens of thousands of people already sustaining over 200 assisted-reproduction clinics in the United States and already availing themselves of in vitro fertilization, intracytoplasmic sperm injection and other techniques of assisted reproduction, cloning would be an option with virtually no added fuss (especially when the success rate improves). Should commercial interests develop in "nucleus-banking," as they have in sperm-banking; should famous athletes or other celebrities decide to market their DNA the way they now market their autographs and just about everything else; should techniques of embryo and germline genetic testing and manipulation arrive as anticipated, increasing the use of laboratory assistance in order to obtain "better" babies—should all this come to pass, then cloning, if it is permitted, could become more than a marginal practice simply on the basis of free reproductive choice, even without any social encouragement to upgrade the gene pool or to replicate superior types. Moreover, if laboratory research on human cloning proceeds, even without any intention to produce cloned humans, the existence of cloned human embryos in the laboratory, created to begin with only for research purposes, would surely pave the way for later baby-making implantations.

In anticipation of human cloning, apologists and proponents have already made clear possible uses of the perfected technology, ranging from the sentimental and compassionate to the grandiose. They include: providing a child for an infertile couple; "replacing" a beloved spouse or child who is dying or has died; avoiding the risk of genetic disease; permitting reproduction for homosexual men and lesbians who want nothing sexual to do with the opposite sex; securing a genetically identical source of organs or tissues perfectly suitable for transplantation; getting a child with a genotype of one's own choosing, not excluding oneself; replicating individuals of great genius, talent, or beauty—having a child who really could "be like Mike"; and creating large sets of genetically identical humans suitable for research on, for instance, the question of nature versus nurture, or for special missions in peace and war (not excluding espionage), in which using identical humans would be an advantage. Most people who envision the cloning of human beings, of course, want none of these scenarios. That they cannot say why is not surprising. What is surprising, and welcome, is that, in our cynical age, they are saying anything at all.

THE WISDOM OF REPUGNANCE

"Offensive." "Grotesque." "Revolting." "Repugnant." "Repulsive." These are the words most commonly heard regarding the prospect of human cloning. Such reactions come both from the man or woman in the street and from the intellectuals, from believers and atheists, from humanists and scientists. Even Dolly's creator has said he "would find it offensive" to clone a human being.

People are repelled by many aspects of human cloning. They recoil from the prospect of mass production of human beings, with large clones of look-alikes, compromised in their individuality; the idea of father-son or mother-daughter twins; the bizarre prospects of a woman giving birth to and rearing a genetic copy of herself, her spouse, or even her deceased father or mother; the grotesqueness of conceiving a child as an exact replacement for another who has died; the utilitarian creation of embryonic genetic duplicates of oneself, to be frozen away or created when necessary, in case of need for homologous tissues or organs for transplantation; the narcissism of those who would clone themselves and the arrogance of others who think they know who deserves to be cloned or which genotype any child-to-be should be thrilled to receive; the Frankensteinian hubris to create human life and increasingly to control its destiny; man playing God. Almost no one finds any of the suggested reasons for human cloning compelling; almost everyone anticipates its possible misuses and abuses. Moreover, many people feel oppressed by the sense that there is probably nothing we can do to prevent it from happening. This makes the prospect all the more revolting.

Revulsion is not an argument; and some of yesterday's repugnances are today calmly accepted—though, one must add, not always for the better. In crucial cases, however, repugnance is the emotional expression of deep wisdom, beyond reason's power fully to articulate it. Can anyone really give an argument fully adequate to the horror which is father-daughter incest (even with consent), or having sex with animals, or mutilating a corpse, or eating human flesh, or even just (just!) raping or murdering another human being? Would anybody's failure to give full rational justification for his or her revulsion at these practices make that revulsion ethically suspect? Not at all. On the contrary, we are suspicious of those who think that they can rationalize away our horror, say, by trying to explain the enormity of incest with arguments only about the genetic risks of inbreeding.

The repugnance at human cloning belongs in this category. We are repelled by the prospect of cloning human beings not because of the strangeness or novelty of the undertaking, but because we intuit and feel, immediately and without argument, the violation of things that we rightfully hold dear. Repugnance, here as elsewhere, revolts against the excesses of human willfulness, warning us not to transgress what is unspeakably profound. Indeed, in this age in which everything is held to be permissible so long as it is freely done, in which our given human nature no longer commands respect, in which our bodies are regarded as mere instruments of our autonomous rational wills, repugnance may be the only voice left that speaks up to defend the central core of our humanity. Shallow are the souls that have forgotten how to shudder.

The goods protected by repugnance are generally overlooked by our customary ways of approaching all new biomedical technologies. The way we evaluate cloning ethically will in fact be shaped by how we characterize it descriptively, by the context into which we place it, and by the perspective from which we view it. The first task for ethics is proper description. And here is where our failure begins.

Typically, cloning is discussed in one or more of three familiar contexts, which one might call the technological, the liberal, and the meliorist. Under the first, cloning will be seen as an extension of existing techniques for assisting reproduction and determining the genetic makeup of children. Like them, cloning is to be regarded as a neutral technique, with no inherent meaning or goodness, but subject to multiple uses, some good, some bad. The morality of cloning thus depends absolutely on the goodness or badness of the motives and intentions of the cloners: as one bioethicist defender of cloning puts it, "the ethics must be judged [only] by the way the parents nurture and rear their resulting child and whether they bestow the same love and affection on a child brought into existence by a technique of assisted reproduction as they would on a child born in the usual way."

The liberal (or libertarian or liberationist) perspective sets cloning in the context of rights, freedoms, and personal empowerment. Cloning is just a new option for exercising an individual's right to reproduce or to have the kind of child that he or she wants. Alternatively, cloning enhances our liberation (especially women's liberation) from the confines of nature, the vagaries of chance, or the necessity for sexual mating. Indeed, it liberates women from the need for men altogether, for the process requires only eggs, nuclei, and (for the time being) uteri—plus, of course, a healthy dose of our (allegedly "masculine") manipulative science that likes to do all these things to mother nature and nature's mothers. For those who hold this outlook, the only moral restraints on cloning are adequately informed consent and the avoidance of bodily harm. If no one is cloned without her consent, and if the clonant is not physically damaged, then the liberal conditions for licit, hence moral, conduct are met. Worries that go beyond violating the will or maiming the body are dismissed as "symbolic"—which is to say, unreal.

The meliorist perspective embraces valetudinarians and also eugenicists. The latter were formerly more vocal in these discussions, but they are now generally happy to see their goals advanced under the less threatening banners of freedom and technological growth. These people see in cloning a new prospect for improving human beings—minimally, by ensuring the perpetuation of healthy individuals by avoiding the risks of genetic disease inherent in the lottery of sex, and maximally, by producing "optimum babies," preserving outstanding genetic material, and (with the help of soon-to-come techniques for precise genetic engineering) enhancing inborn human capacities on many fronts. Here the morality of cloning as a means is justified solely by the excellence of the end, that is, by the outstanding traits or individuals cloned—beauty, or brawn, or brains.

These three approaches, all quintessentially American and all perfectly fine in their places, are sorely wanting as approaches to human procreation. It is, to say the least, grossly distorting to view the wondrous mysteries of birth, renewal, and individuality, and the deep meaning of parent-child relations, largely through the lens of our reductive science and its potent technologies. Similarly, considering reproduction (and the intimate relations of family life!) primarily under the political-legal, adversarial, and individualistic notion of rights can only undermine the private yet fundamentally social, cooperative, and duty-laden character of childbearing, child rearing, and their bond to the covenant of marriage. Seeking to escape entirely from nature (in order to satisfy a natural desire or a natural right to reproduce!) is self-contradictory in theory and self-alienating in practice. For we are erotic beings only because we are embodied beings, and not merely intellects and wills unfortunately imprisoned in our bodies. And, though health and fitness are clearly great goods, there is something deeply disquieting in looking on our prospective children as artful products perfectible by genetic engineering, increasingly held to our willfully imposed designs, specifications, and margins of tolerable error.

The technical, liberal, and meliorist approaches all ignore the deeper anthropological, social, and, indeed, ontological meanings of bringing forth new life. To this more fitting and profound point of view, cloning shows itself to be a major alteration, indeed, a major violation, of our given nature as embodied, gendered, and engendering beings—and of the social relations built on this natural ground. Once this perspective is recognized, the ethical judgment on cloning can no longer be reduced to a matter of motives and intentions, rights and freedoms, benefits and harms, or even means and ends. It must be regarded primarily as a matter of meaning: Is cloning a fulfillment of human begetting and belonging? Or is cloning rather, as I contend, their pollution and perversion? To pollution and perversion, the fitting response can only be horror and revulsion; and conversely, generalized horror and revulsion are prima facie evidence of foulness and violation. The burden of moral argument must fall entirely on those who want to declare the widespread repugnances of humankind to be mere timidity or superstition.

Yet repugnance need not stand naked before the bar of reason. The wisdom of our horror at human cloning can be partially articulated, even if this is finally one of those instances about which the heart has its reasons that reason cannot entirely know.

THE PROFUNDITY OF SEX

To see cloning in its proper context, we must begin not, as I did before, with laboratory technique, but with the anthropology—natural and social—of sexual reproduction.

Sexual reproduction—by which I mean the generation of new life from (exactly) two complementary elements, one female, one male, (usually) through coitus—is established (if that is the right term) not by human decision, culture or tradition, but by nature; it is the natural way of all mammalian reproduction. By nature, each child has two complementary biological progenitors. Each child thus stems from and unites exactly two lineages. In natural generation, moreover, the precise genetic constitution of the resulting offspring is determined by a combination of nature and chance, not by human design: each human child shares the common natural human species genotype, each child is genetically (equally) kin to each (both) parent(s), yet each child is also genetically unique.

These biological truths about our origins foretell deep truths about our identity and about our human condition altogether. Every one of us is at once equally human, equally enmeshed in a particular familial nexus of origin, and equally individuated in our trajectory from birth to death—and, if all goes well, equally capable (despite our morality) of participating, with a complementary other, in the very same renewal of such human possibility through procreation. Though less momentous than our common humanity, our genetic individuality is not humanly trivial. It shows itself forth in our distinctive appearance through which we are everywhere recognized; it is revealed in our "signature" marks of fingerprints and our self-recognizing immune system; it symbolizes and foreshadows exactly the unique, never-to-be-repeated character of each human life.

Human societies virtually everywhere have structured child-rearing responsibilities and systems of identity and relationship on the bases of these deep natural facts of begetting. The mysterious yet ubiquitous "love of one's own" is everywhere culturally exploited, to make sure that children are not just produced but well cared for and to create for everyone clear ties of meaning, belonging, and obligation. But it is wrong to treat such naturally rooted social practices as mere cultural constructs (like left- or right-driving, or like burying or cremating the dead) that we can alter with little human cost. What would kinship be without its clear natural grounding? And what would identity be without kinship? We must resist those who have begun to refer to sexual reproduction as the "traditional method of reproduction," who would have us regard as merely traditional, and by implication arbitrary, what is in truth not only natural but most certainly profound.

Asexual reproduction, which produces "single-parent" offspring, is a radical departure from the natural human way, confounding all normal understandings of father, mother, sibling, grandparent, etc., and all moral relations tied thereto. It becomes even more of a radical departure when the resulting offspring is a clone derived not from an embryo, but from a mature adult to whom the clone would be an identical twin; and when the process occurs not by natural accident (as in natural twinning), but by deliberate human design and manipulation; and when the

CLONING

child's (or children's) genetic constitution is preselected by the parent(s) (or scientists). Accordingly, as we will see, cloning is vulnerable to three kinds of concerns and objections, related to these three points: cloning threatens confusion of identity and individuality, even in small-scale cloning; cloning represents a giant step (though not the first one) toward transforming procreation into manufacture, that is, toward the increasing depersonalization of the process of generation and, increasingly, toward the "production" of human children as artifacts, products of human will and design (what others have called the problem of "commodification" of new life); and cloning—like other forms of eugenic engineering of the next generation—represents a form of despotism of the cloners over the cloned, and thus (even in benevolent cases) represents a blatant violation of the inner meaning of parent-child relations, of what it means to have a child, of what it means to say "yes" to our own demise and "replacement."

Before turning to these specific ethical objections, let me test my claim of the profundity of the natural way by taking up a challenge recently posed by a friend. What if the given natural human way of reproduction were asexual, and we now had to deal with a new technological innovation—artificially induced sexual dimorphism and the fusing of complementary gametes—whose inventors argued that sexual reproduction promised all sorts of advantages, including hybrid vigor and the creation of greatly increased individuality? Would one then be forced to defend natural asexuality because it was natural? Could one claim that it carried deep human meaning?

The response to this challenge broaches the ontological meaning of sexual reproduction. For it is impossible, I submit, for there to have been human life—or even higher forms of animal life—in the absence of sexuality and sexual reproduction. We find asexual reproduction only in the lowest forms of life: bacteria, algae, fungi, some lower invertebrates. Sexuality brings with it a new and enriched relationship to the world. Only sexual animals can seek and find complementary others with whom to pursue a goal that transcends their own existence. For a sexual being, the world is no longer an indifferent and largely homogeneous *otherness*, in part edible, in part dangerous. It also contains some very special and related and complementary beings, of the same kind but of opposite sex, toward whom one reaches out with special interest and intensity. In higher birds and mammals, the outward gaze keeps a lookout not only for food and predators, but also for prospective mates; the beholding of the many-splendored world is suffused with desire for union, the animal antecedent of human eros and the germ of sociality. Not by accident is the human animal both the sexiest animal—whose females do not go into heat but are receptive throughout the estrous cycle and whose males must therefore have greater sexual appetite and energy in order to reproduce successfully—and also the most aspiring, the most social, the most open, and the most intelligent animal.

The soul-elevating power of sexuality is, at bottom, rooted in its strange connection to mortality, which it simultaneously accepts and tries to overcome. Asexual reproduction may be seen as a continuation of the activity of self-preservation. When one organism buds or divides to become two, the original being is (doubly) preserved, and nothing dies. Sexuality, by contrast, means perishability and serves replacement; the two that come together to generate one soon will die. Sexual desire, in human beings as in animals, thus serves an end that is partly hidden from, and finally at odds with, the self-serving individual. Whether we know it or not, when we are sexually active we are voting with our genitalia for our own demise. The salmon swimming upstream to spawn and die tell the universal story: sex is bound up with death, to which it holds a partial answer in procreation.

The salmon and the other animals evince this truth blindly. Only the human being can understand what it means. As we learn so powerfully from the story of the Garden of Eden, our humanization is coincident with sexual self-consciousness, with the recognition of our sexual nakedness and all that it implies: shame at our needy incompleteness, unruly self-division, and finitude; awe before the eternal; hope in the self-transcending possibilities of children and a relationship to the divine. In the sexually self-conscious animal, sexual desire can become eros, lust can become love. Sexual desire humanly regarded is thus sublimated into erotic longing for wholeness, completion, and immortality, which drives us knowingly into the embrace and its generative fruit—as well as into all the higher human possibilities of deed, speech, and song.

Through children, a good common to both husband and wife, male and female achieve some genuine unification (beyond the mere sexual "union," which fails to do so). The two become one through sharing generous (not needy) love for this third being as good. Flesh of their flesh, the child is the parents' own commingled being externalized, and given a separate and persisting existence. Unification is enhanced also by their commingled work of rearing. Providing an opening to the future beyond the grave, carrying not only our seed but also our names, our ways, and our hopes that they will surpass us in goodness and happiness, children are a testament to the possibility of transcendence. Gender duality and sexual desire, which first draws our love upward and outside of ourselves, finally provide for the partial overcoming of the confinement and limitation of perishable embodiment altogether.

Human procreation, in sum, is not simply an activity of our rational wills. It is a more complete activity precisely because it engages us bodily, erotically, and spiritually, as well as rationally. There is wisdom in the mystery of nature that has joined the pleasure of sex, the inarticulate longing for union, the communication of the loving embrace, and the deep-seated and only partly articulate desire for children in the very activity by which we continue the chain of human existence and participate in the renewal of human possibility. Whether or not we know it, the severing of procreation from sex, love, and intimacy is inherently dehumanizing, no matter how good the product.

We are now ready for the more specific objections to cloning.

The Perversities of Cloning

First, an important if formal objection: any attempt to clone a human being would constitute an unethical experiment upon the resulting child-to-be. As the animal experiments (frog and sheep) indicate, there are grave risks of mishaps and deformities. Moreover, because of what cloning means, one cannot presume a future cloned child's consent to be a clone, even a healthy one. Thus, ethically speaking, we cannot even get to know whether or not human cloning is feasible.

I understand, of course, the philosophical difficulty of trying to compare a life with defects against nonexistence. Several bioethicists, proud of their philosophical cleverness, use this conundrum to embarrass claims that one can injure a child in its conception, precisely because it is only thanks to that complained-of conception that the child is alive to complain. But common sense tells us that we have no reason to fear such philosophisms. For we surely know that people can harm and even maim children in the very act of conceiving them, say, by paternal transmission of the AIDS virus, maternal transmission of heroin dependence, or, arguably, even by bringing them into being as bastards or with no capacity or willingness to look after them properly. And we believe that to do this intentionally, or even negligently, is inexcusable and clearly unethical.

The objection about the impossibility of presuming consent may even go beyond the obvious and sufficient point that a clonant, were he subsequently to be asked, could lightly resent having been made a clone. At issue are not just benefits and harms, but doubts about the very independence needed to give proper (even retroactive) consent, that is, not just the capacity to choose but the disposition and ability to choose freely and well. It is not at all clear to what extent a clone will truly be a moral agent. For, as we shall see, in the very fact of cloning, and of rearing him as a clone, his makers subvert the cloned child's independence, beginning with that aspect that comes from knowing that one was an unbidden surprise, a gift, to the world, rather than the designed result of someone's artful project.

Cloning creates serious issues of identity and individuality. The cloned person may experience concerns about his distinctive identity not only because he will be in genotype and appearance identical to another human being, but, in this case, because he may also be twin to the person who is his "father" or "mother"—if one can still call them that. What would be the psychic burdens of being the "child" or "parent" of your twin? The cloned individual, moreover, will be saddled with a genotype that has already lived. He will not be fully a surprise to the world. People are likely always to compare his performances in life with that of his alter ego. True, his nurture and his circumstance in life will be different; genotype is not exactly destiny. Still, one must also expect parental and other efforts to shape this new life after the original—or at least to view the child with the original version always firmly in mind. Why else did they clone from the star basketball player, mathematician, and beauty queen—or even dear old dad—in the first place?

Since the birth of Dolly, there has been a fair amount of doublespeak on this matter of genetic identity. Experts have rushed in to reassure the public that the clone would in no way be the same person, or have any confusions about his or her identity: as previously noted, they are pleased to point out that the clone of Mel Gibson would not be Mel Gibson. Fair enough. But one is shortchanging the truth by emphasizing the additional importance of the intrauterine environment, rearing, and social setting: genotype obviously matters plenty. That, after all, is the only reason to clone, whether human beings or sheep. The odds that clones of Wilt Chamberlain will play in the NBA are, I submit, infinitely greater than they are for clones of Robert Reich.

Curiously, this conclusion is supported, inadvertently, by the one ethical sticking point insisted on by friends of cloning: no cloning without the donor's consent. Though an orthodox liberal objection, it is in fact quite puzzling when it comes from people (such as Ruth Macklin) who also insist that genotype is not identity or individuality, and who deny that a child could reasonably complain about being made a genetic copy. If the clone of Mel Gibson would not be Mel Gibson, why should Mel Gibson have grounds to object that someone had been made his clone? We already allow researchers to use blood and tissue samples for research purposes of no benefit to their sources: my falling hair, my expectorations, my urine, and even my biopsied tissues are "not me" and not mine. Courts have held that the profit gained from uses to which scientists put my discarded tissues do not legally belong to me. Why, then, no cloning without consent—including, I assume, no cloning from the body of someone who just died? What harm is done the donor, if genotype is "not me"? Truth to tell, the only powerful justification for objecting is that genotype really does have something to do with identity, and everybody knows it. If not, on what basis could Michael Jordan object that someone cloned "him," say, from cells taken from a "lost" scraped-off piece of his skin? The insistence on donor consent unwittingly reveals the problem of identity in all cloning.

Genetic distinctiveness not only symbolizes the uniqueness of each human life and the independence of its parents that each human child rightfully attains. It can also be an important support for living a worthy and dignified life. Such arguments apply with great force to any large-scale replication of human individuals. But they are sufficient, in my view, to rebut even the first attempts to clone a human being. One must never forget that these are human beings upon whom our eugenic or merely playful fantasies are to be enacted.

Troubled psychic identity (distinctiveness), based on all-too-evident genetic identity (sameness), will be made much worse by the utter confusion of social identity and kinship ties. For, as already noted, cloning radically confounds lineage and social relations, for "offspring" as for "parents." As bioethicist James Nelson has pointed out, a female child cloned from her "mother" might develop a desire for a relationship to her "father," and might understandably seek out the father of her "mother," who is after all also her biological twin sister. Would "grandpa," who thought his paternal duties concluded, be pleased to discover that the clonant looked to him for paternal attention and support?

Social identity and social ties of relationship and responsibility are widely connected to, and supported by, biological kinship. Social taboos on incest (and adultery) everywhere serve to keep clear who is related to whom (and especially which child belongs to which parents), as well as to avoid confounding the social identity of parent-and-child (or brother-and-sister) with the social identity of lovers, spouses, and co-parents. True, social identity is altered by adoption (but as a matter of the best interest of already living children: we do not deliberately produce children for adoption). True, artificial insemination and in vitro fertilization with donor sperm, or whole embryo donation, are in some way forms of "prenatal adoption"— a not altogether unproblematic practice. Even here, though, there is in each case (as in all sexual reproduction) a known male source of sperm and a known single female source of egg—a genetic father and a genetic mother—should anyone care to know (as adopted children often do) who is genetically related to whom.

In the case of cloning, however, there is but one "parent." The usually sad situation of the "single-parent child" is here deliberately planned, and with a vengeance. In the case of self-cloning, the "offspring" is, in addition, one's twin; and so the dreaded result of incest—to be parent to one's sibling—is here brought about deliberately, albeit without any act of coitus. Moreover, all other relationships will be confounded. What will father, grandfather, aunt, cousin, sister mean? Who will bear what ties and what burdens? What sort of social identity will someone have with one whole side—"father's" or "mother's"—necessarily excluded? It is no answer to say that our society, with its high incidence of divorce, remarriage, adoption, extramarital childbearing, and the rest, already confounds lineage and confuses kinship and responsibility for children (and everyone else), unless one also wants to argue that this is, for children, a preferable state of affairs.

Human cloning would also represent a giant step toward turning begetting into making, procreation into manufacture (literally, something "handmade"), a process already begun with in vitro fertilization and genetic testing of embryos. With cloning, not only is the process in hand, but the total genetic blueprint of the cloned individual is selected and determined by the human artisans. To be sure, subsequent development will take place according to natural processes; and the resulting children will still be recognizably human. But we here would be taking a major step into making man himself simply another one of the man-made things. Human nature becomes merely the last part of nature to succumb to the technological project, which turns all of nature into raw material at human disposal, to be homogenized by our rationalized technique according to the subjective prejudices of the day.

How does begetting differ from making? In natural procreation, human beings come together, complementarily male and female, to give existence to another being who is formed, exactly as we were, *by what we are*: living, hence perishable, hence aspiringly erotic, human beings. In clonal reproduction, by contrast, and in the more advanced forms of manufacture to which it leads, we give existence to a being not by what we are but by what we intend and design. As with any product of our making, no matter how excellent, the artificer stands above it, not as an equal but as a superior, transcending it by his will and creative prowess. Scientists who clone animals make it perfectly clear that they are engaged in instrumental making; the animals are, from the start, designed as means to serve rational human purposes. In human cloning, scientists and prospective "parents" would be adopting the same technocratic mentality to human children: human children would be their artifacts.

Such an arrangement is profoundly dehumanizing, no matter how good the product. Mass-scale cloning of the same individual makes the point vividly; but the violation of human equality, freedom, and dignity are present even in a single planned clone. And procreation dehumanized into manufacture is further degraded by commodification, a virtually inescapable result of allowing baby making to proceed under the banner of commerce. Genetic and reproductive biotechnology companies are already growth industries, but they will go into commercial orbit once the Human Genome Project nears completion. Supply will create enormous demand. Even before the capacity for human cloning arrives, established companies will have invested in the harvesting of eggs from ovaries

CLONING

obtained at autopsy or through ovarian surgery, practiced embryonic genetic alteration, and initiated the stockpiling of prospective donor tissues. Through the
rental of surrogate-womb services, and through the buying and selling of tissues
and embryos, priced according to the merit of the donor, the commodification of
nascent human life will be unstoppable.

Finally, and perhaps most important, the practice of human cloning by
nuclear transfer—like other anticipated forms of genetic engineering of the next
generation—would enshrine and aggravate a profound and mischievous misunderstanding of the meaning of having children and of the parent-child relationship. When a couple now chooses to procreate, the partners are saying yes to the
emergence of new life in its novelty, saying yes not only to having a child but also,
tacitly, to having whatever child this child turns out to be. In accepting our finitude
and opening ourselves to our replacement, we are tacitly confessing the limits of
our control. In this ubiquitous way of nature, embracing the future by procreating
means precisely that we are relinquishing our grip, in the very activity of taking
up our own share in what we hope will be the immortality of human life and the
human species. This means that our children are not *our* children: they are not our
property, not our possessions. Neither are they supposed to live our lives for us, or
anyone else's life but their own. To be sure, we seek to guide them on their way,
imparting to them not just life but nurturing, love, and a way of life; to be sure,
they bear our hopes that they will live fine and flourishing lives, enabling us in
small measure to transcend our own limitations. Still, their genetic distinctiveness
and independence are the natural foreshadowing of the deep truth that they have
their own and never-before-enacted life to live. They are sprung from a past, but
they take an uncharted course into the future.

Much harm is already done by parents who try to live vicariously through
their children. Children are sometimes compelled to fulfill the broken dreams of
unhappy parents; John Doe Jr. or the III is under the burden of having to live up
to his forebear's name. Still, if most parents have hopes for their children, cloning
parents will have expectations. In cloning, such overbearing parents take at the
start a decisive step which contradicts the entire meaning of the open and for-
ward-looking nature of parent-child relations. The child is given a genotype that
has already lived, with full expectation that this blueprint of a past life ought to
be controlling of the life that is to come. Cloning is inherently despotic, for it
seeks to make one's children (or someone else's children) after one's own image (or
an image of one's choosing) and their future according to one's will. In some cases,
the despotism may be mild and benevolent. In other cases, it will be mischievous
and downright tyrannical. But despotism—the control of another through one's
will—it inevitably will be.

MEETING SOME OBJECTIONS

The defenders of cloning, of course, are not wittingly friends of despotism. Indeed, they regard themselves mainly as friends of freedom: the freedom of individuals to reproduce, the freedom of scientists and inventors to discover and devise and to foster "progress" in genetic knowledge and technique. They want large-scale cloning only for animals, but they wish to preserve cloning as a human option for exercising our "right to reproduce"—our right to have children, and children with "desirable genes." As law professor John Robertson points out, under our "right to reproduce" we already practice early forms of unnatural, artificial, and extramarital reproduction, and we already practice early forms of eugenic choice. For this reason, he argues, cloning is no big deal.

We have here a perfect example of the logic of the slippery slope, and the slippery way in which it already works in this area. Only a few years ago, slippery slope arguments were used to oppose artificial insemination and in vitro fertilization using unrelated sperm donors. Principles used to justify these practices, it was said, will be used to justify more artificial and more eugenic practices, including cloning. Not so, the defenders retorted, since we can make the necessary distinctions. And now, without even a gesture at making the necessary distinctions, the continuity of practice is held by itself to be justificatory.

The principle of reproductive freedom as currently enunciated by the proponents of cloning logically embraces the ethical acceptability of sliding down the entire rest of the slope—to producing children ectogenetically from sperm to term (should it become feasible) and to producing children whose entire genetic makeup will be the product of parental eugenic planning and choice. If reproductive freedom means the right to have a child of one's own choosing, by whatever means, it knows and accepts no limits.

But, far from being legitimated by a "right to reproduce," the emergence of techniques of assisted reproduction and genetic engineering should compel us to reconsider the meaning and limits of such a putative right. In truth, a "right to reproduce" has always been a peculiar and problematic notion. Rights generally belong to individuals, but this is a right which (before cloning) no one can exercise alone. Does the right then inhere only in couples? Only in married couples? Is it a (woman's) right to carry or deliver or a right (of one or more parents) to nurture and rear? Is it a right to have your own biological child? Is it a right only to attempt reproduction, or a right also to succeed? Is it a right to acquire the baby of one's choice?

The assertion of a negative "right to reproduce" certainly makes sense when it claims protection against state interference with procreative liberty, say, through a program of compulsory sterilization. But surely it cannot be the basis of a tort claim against nature, to be made good by technology, should free efforts at natural procreation fail. Some insist that the right to reproduce embraces also the right against state interference with the free use of all technological means to obtain a child. Yet such a position cannot be sustained: for reasons having to do with the means employed, any community may rightfully prohibit surrogate pregnancy, or polygamy, or the sale of babies to infertile couples, without violating anyone's basic human "right to reproduce." When the exercise of a previously innocuous freedom now involves or impinges on troublesome practices that the original freedom never was intended to reach, the general presumption of liberty needs to be reconsidered.

We do indeed already practice negative eugenic selection, through genetic screening and prenatal diagnosis. Yet our practices are governed by a norm of health. We seek to prevent the birth of children who suffer from known (serious) genetic diseases. When and if gene therapy becomes possible, such diseases could then be treated, in utero or even before implantation—I have no ethical objection in principle to such a practice (though I have some practical worries), precisely because it serves the medical goal of healing existing individuals. But therapy, to be therapy, implies not only an existing "patient." It also implies a norm of health. In this respect, even germline gene "therapy," though practiced not on a human being but on egg and sperm, is less radical than cloning, which is in no way therapeutic. But once one blurs the distinction between health promotion and genetic enhancement, between so-called negative and positive eugenics, one opens the door to all future eugenic designs. "To make sure that a child will be healthy and have good chances in life": this is Robertson's principle, and owing to its latter clause it is an utterly elastic principle, with no boundaries. Being over eight feet tall will likely produce some very good chances in life, and so will having the looks of Marilyn Monroe, and so will a genius-level intelligence.

Proponents want us to believe that there are legitimate uses of cloning that can be distinguished from illegitimate uses, but by their own principles no such limits can be found. (Nor could any such limits be enforced in practice.) Reproductive freedom, as they understand it, is governed solely by the subjective wishes of the parents-to-be (plus the avoidance of bodily harm to the child). The sentimentally appealing case of the childless married couple is, on these grounds, indistinguishable from the case of an individual (married or not) who would like to clone someone famous or talented, living or dead. Further, the principle here endorsed justifies not only cloning but, indeed, all future artificial attempts to create (manufacture) "perfect" babies.

A concrete example will show how, in practice no less than in principle, the so-called innocent case will merge with, or even turn into, the more troubling ones. In practice, the eager parents-to-be will necessarily be subject to the tyranny of expertise. Consider an infertile married couple, she lacking eggs or he lacking sperm, that wants a child of their (genetic) own, and propose to clone either husband or wife. The scientist-physician (who is also co-owner of the cloning company) points out the likely difficulties—a cloned child is not really their (genetic) child, but the child of only *one* of them; this imbalance may produce strains on the marriage; the child might suffer identity confusion; there is a risk of perpetuating the cause of sterility; and so on—and he also points out the advantages of choosing a donor nucleus. Far better than a child of their own would be a child of their own choosing. Touting his own expertise in selecting healthy and talented donors, the doctor presents the couple with his latest catalog containing the pictures, the health records, and the accomplishments of his stable of cloning donors, samples of whose tissues are in his deep freeze. Why not, dearly beloved, a more perfect baby?

The "perfect baby," of course, is the project not of the infertility doctors, but of the eugenic scientists and their supporters. For them, the paramount right is not the so-called right to reproduce but what biologist Bentley Glass called, a quarter of a century ago, "the right of every child to be born with a sound physical and mental constitution, based on a sound genotype ... the inalienable right to a sound heritage." But to secure this right, and to achieve the requisite quality control over new human life, human conception and gestation will need to be brought fully into the bright light of the laboratory, beneath which it can be fertilized, nourished, pruned, weeded, watched, inspected, prodded, pinched, cajoled, injected, tested, rated, graded, approved, stamped, wrapped, sealed, and delivered. There is no other way to produce the perfect baby.

Yet we are urged by proponents of cloning to forget about the science-fiction scenarios of laboratory manufacture and multiple-copied clones, and to focus only on the homely cases of infertile couples exercising their reproductive rights. But why, if the single cases are so innocent, should multiplying their performance be so off-putting? (Similarly, why do others object to people making money off this practice, if the practice itself is perfectly acceptable?) When we follow the sound ethical principle of universalizing our choice—"would it be right if everyone cloned a Wilt Chamberlain (with his consent, of course)? Would it be right if everyone decided to practice asexual reproduction?"—we discover what is wrong with these seemingly innocent cases. The so-called science-fiction cases make vivid the meaning of what looks to us, mistakenly, to be benign.

Though I recognize certain continuities between cloning and, say, in vitro fertilization, I believe that cloning differs in essential and important ways. Yet those who disagree should be reminded that the "continuity" argument cuts both ways. Sometimes we establish bad precedents, and discover that they were bad only when we follow their inexorable logic to places we never meant to go. Can the defenders of cloning show us today how, on their principles, we will be able to see producing babies ("perfect babies") entirely in the laboratory or exercising full control over their genotypes (including so-called enhancement) as ethically different, in any essential way, from present forms of assisted reproduction? Or are they willing to admit, despite their attachment to the principle of continuity, that the complete obliteration of "mother" or "father," the complete depersonalization of procreation, the complete manufacture of human beings, and the complete genetic control of one generation over the next would be ethically problematic and essentially different from current forms of assisted reproduction? If so, where and how will they draw the line, and why? I draw it at cloning, for all the reasons given.

Ban the Cloning of Humans

What, then, should we do? We should declare that human cloning is unethical in itself and dangerous in its likely consequences. In so doing, we shall have the backing of the overwhelming majority of our fellow Americans, and of the human race, and (I believe) of most practicing scientists. Next, we should do all that we can to prevent the cloning of human beings. We should do this by means of an international legal ban if possible, and by a unilateral national ban, at a minimum. Scientists may secretly undertake to violate such a law, but they will be deterred by not being able to stand up proudly to claim the credit for their technological bravado and success. Such a ban on clonal baby making, moreover, will not harm the progress of basic genetic science and technology. On the contrary, it will reassure the public that scientists are happy to proceed without violating the deep ethical norms and intuitions of the human community.

This still leaves the vexed question about laboratory research using early embryonic human clones, specially created only for such research purposes, with no intention to implant them into a uterus. There is no question that such research holds great promise for gaining fundamental knowledge about normal (and abnormal) differentiation, and for developing tissue lines for transplantation that might be used, say, in treating leukemia or in repairing brain or spinal cord injuries—to mention just a few of the conceivable benefits. Still, unrestricted clonal embryo research will surely make the production of living human clones much more likely. Once the genies put the cloned embryos into the bottles, who can strictly control where they go (especially in the absence of legal prohibitions against implanting them to produce a child)?

I appreciate the potentially great gains in scientific knowledge and medical treatment available from embryo research, especially with cloned embryos. At the same time, I have serious reservations about creating human embryos for the sole purpose of experimentation. There is something deeply repugnant and fundamentally transgressive about such a utilitarian treatment of prospective human life. This total, shameless exploitation is worse, in my opinion, than the "mere" destruction of nascent life. But I see no added objections, as a matter of principle, to creating and using *cloned* early embryos for research purposes, beyond the objections that I might raise to doing so with embryos produced sexually.

And yet, as a matter of policy and prudence, any opponent of the manufacture of cloned humans must, I think, in the end oppose also the creating of cloned human embryos. Frozen embryonic clones (belonging to whom?) can be shuttled around without detection. Commercial ventures in human cloning will be developed without adequate oversight. In order to build a fence around the law, prudence dictates that one oppose—for this reason alone—all production of cloned human embryos, even for research purposes. We should allow all cloning research on animals to go forward, but the only safe trench that we can dig across the slippery slope, I suspect, is to insist on the inviolable distinction between animal and human cloning.

Some readers, and certainly most scientists, will not accept such prudent restraints, since they desire the benefits of research. They will prefer, even in fear and trembling, to allow human embryo cloning research to go forward.

Very well. Let us test them. If the scientists want to be taken seriously on ethical grounds, they must at the very least agree that embryonic research may proceed if and only if it is preceded by an absolute and effective ban on all attempts to implant into a uterus a cloned human embryo (cloned from an adult) to produce a living child. Absolutely no permission for the former without the latter.

The National Bioethics Advisory Commission's recommendations regarding this matter should be watched with the greatest care. Yielding to the wishes of the scientists, the commission will almost surely recommend that cloning human embryos for research be permitted. To allay public concern, it will likely also call for a temporary moratorium—not a legislative ban—on implanting cloned embryos to make a child, at least until such time as cloning techniques will have been perfected and rendered "safe" (precisely through the permitted research with cloned embryos). But the call for a moratorium rather than a legal ban would be a moral and a practical failure. Morally, this ethics commission would (at best) be waffling on the main ethical question, by refusing to declare the production of human clones unethical (or ethical). Practically, a moratorium on implantation cannot provide even the minimum protection needed to prevent the production of cloned humans.

CLONING

Opponents of cloning need therefore to be vigilant. Indeed, no one should be willing even to consider a recommendation to allow the embryo research to proceed unless it is accompanied by a call for *prohibiting* implantation and until steps are taken to make such a prohibition effective.

Technically, the National Bioethics Advisory Commission can advise the president only on federal policy, especially federal funding policy. But given the seriousness of the matter at hand, and the grave public concern that goes beyond federal funding, the commission should take a broader view. (If it doesn't, Congress surely will.) Given that most assisted reproduction occurs in the private sector, it would be cowardly and insufficient for the commission to say, simply, "no federal funding" for such practices. It would be disingenuous to argue that we should allow federal funding so that we would then be able to regulate the practice; the private sector will not be bound by such regulations. Far better, for virtually everyone concerned, would be to distinguish between research on embryos and baby making, and to call for a complete national and international ban (effected by legislation and treaty) of the latter, while allowing the former to proceed (at least in private laboratories).

The proposal for such a legislative ban is without American precedent, at least in technological matters, though the British and others have banned cloning of human beings, and we ourselves ban incest, polygamy, and other forms of "reproductive freedom." Needless to say, working out the details of such a ban, especially a global one, would be tricky, what with the need to develop appropriate sanctions for violators. Perhaps such a ban will prove ineffective; perhaps it will eventually be shown to have been a mistake. But it would at least place the burden of practical proof where it belongs: on the proponents of this horror, requiring them to show very clearly what great social or medical good can be had only by the cloning of human beings.

We Americans have lived by, and prospered under, a rosy optimism about scientific and technological progress. The technological imperative—if it can be done, it must be done—has probably served us well, though we should admit that there is no accurate method for weighing benefits and harms. Even when, as in the cases of environmental pollution, urban decay or the lingering deaths that are the unintended by-products of medical success, we recognize the unwelcome outcomes of technological advance, we remain confident in our ability to fix all the "bad" consequences—usually by means of still newer and better technologies. How successfully we can continue to be in such post hoc repairing is at least an open question. But there is very good reason for shifting the paradigm around, at least regarding those technological interventions into the human body and mind that will surely effect fundamental (and likely irreversible) changes in human nature, basic human relationships, and what it means to be a human being. Here we surely should not be willing to risk everything in the naive hope that, should things go wrong, we can later set them right.

The president's call for a moratorium on human cloning has given us an important opportunity. In a truly unprecedented way, we can strike a blow for the human control of the technological project, for wisdom, prudence, and human dignity. The prospect of human cloning, so repulsive to contemplate, is the occasion for deciding whether we shall be slaves of unregulated progress, and ultimately its artifacts, or whether we shall remain free human beings who guide our technique toward the enhancement of human dignity. If we are to seize the occasion, we must, as the late Paul Ramsey wrote,

> raise the ethical questions with a serious and not a frivolous conscience. A man of frivolous conscience announces that there are ethical quandaries ahead that we must urgently consider before the future catches up with us. By this he often means that we need to devise a new ethics that will provide the rationalization for doing in the future what men are bound to do because of new actions and interventions science will have made possible. In contrast a man of serious conscience means to say in raising urgent ethical questions that there may be some things that men should never do. The good things that men do can be made complete only by the things they refuse to do.

HUMAN CLONING: THE CASE FOR

Now it is the turn of those who think that one can make a positive, moral case for human cloning. Walter Glannon draws a distinction between cloning for parts, for instance, for body tissues, and cloning for full human beings. By and large he has little trouble with the former, thinking it is justified morally and an extension of what we already consider right and proper. He has more trouble with the cloning of people as such, although having examined matters from a Kantian perspective, he feels that there are times when a case can be made for such a practice. Suppose one had a loved child who was dying of leukemia and one wanted to clone him or her, to produce a new child without the leukemia who could then be used to help the afflicted first child. Obviously the new child has been produced as a means, but Glannon stresses not necessarily as a means alone. The cloned child could well be loved as an end in him or herself. Here, therefore, Glannon finds no Kantian prohibition.

Glannon also considers whether permitting any cloning whatsoever, for however good a reason, is the slippery slope to some form of eugenics. Are we not on the way to attempting to produce perfect humans according to some predetermined criterion of excellence, rather than simply helping people who are sick or otherwise in need? Although he recognizes that there are dangers here, Glannon does not find them so overwhelming as to warrant prohibiting cloning entirely. Such a conclusion is shared by John Harris. He takes strong issue with some of those whose work we have already encountered, especially the arguments about individuality by Axel Kahn. While appreciative of the significance of human dignity, Harris finds the arguments somewhat unhelpful in deciding moral issues. Kantianism he believes to be "seldom helpful in medical or bioscience contexts." Blood transfusions, Harris argues, are clearly morally worthwhile things but seem not to involve treating folk as ends rather than means. Morally more profitable would be a milder imperative, such as it is better to do some good rather than no good. On such a principle as this, one can make a case for some forms of cloning. "I cannot but think that if it is right to use embryos for research or therapy then it is also right to produce them for such purposes."

More than this, Harris argues that it may even be morally permissible to clone viable humans. At least, he feels that preventing people from having the option of cloning themselves or others might be to produce a greater evil. Opponents of cloning insist on the primacy of human dignity. A case might be made for saying that dignity is better preserved by allowing people to make their own choices rather than having them imposed by others. In Harris's opinion, therefore, we have no good grounds now for calling for an immediate and total ban on human cloning.

(13) The Ethics of Human Cloning
Walter Glannon

I. INTRODUCTION

T
he finding by Ian Wilmut and his colleagues that viable lambs could be
produced by transplanting cell nuclei from adult sheep embryos to enucleated eggs has raised the possibility of cloning human cells to grow tissues and
organs for transplantation.[1] It even has raised the possibility of cloning complete
human organisms. The fascination surrounding the prospect of cloning human
beings has been tempered with caution, however. Indeed, on the recommendations
of the National Bioethics Advisory Committee last June President Clinton
drafted the Cloning Prohibition Act of 1997, which outlaws somatic-cell nuclear
transfer for the purpose of creating a human being. Significantly, the act does not
call for an outright ban on cloning research in general, but includes a five-year
"sunset clause" that allows important and promising work to clone molecules,
DNA, cells, tissues, or animals.[2]

Reprinted from *Public Affairs Quarterly* 12, no. 3 (July 1998): 287–305. Reprinted with permission.

CLONING

The rationale for at least a temporary prohibition on cloning humans should be readily apparent. It is not yet known whether the technique would be safe for the individuals resulting from it. At a deeper level, cloning raises fundamental ethical questions about what makes us human beings, and some have argued that cloning is morally repugnant precisely because it involves a form of biological manipulation which violates what are presumed to be inviolable essential features of our humanity.[3] While cloning arguably is in many respects an extension of existing procedures of assisted reproduction, such as artificial insemination and in vitro fertilization, the production of genetically identical copies of ourselves seems anathema to our deep-seated conviction about the uniqueness of each person.

Possible uses of cloning technology in humans include: (1) "replacing" a loved one who is dying or has died; (2) producing an individual with specially designed physical or cognitive traits; (3) providing a child for an infertile couple, when all other reproductive alternatives have been exhausted; (4) producing a child for a lesbian or gay couple, (5) preventing genetic diseases in people by genetic intervention at the embryonic stage of development; and (6) growing tissues or organs for transplantation. Intuitively, these possibilities fall along a rough continuum of moral justification, with (5) and (6) being perhaps the most justifiable use and (1) and (2) the least. For while (5) and (6) are the most likely to prevent harm to persons by preventing or curing disease, (1) and (2) are the most likely to cause harm to persons by treating them as mere means and not also as ends-in-themselves. I will flesh out these intuitions by exploring the ethical implications of each of these uses. After first addressing the biological aspects of cloning, I will spell out (the grounds on which (1)–(6) can or cannot be morally justified. In examining these issues, I will be guided mainly by the Kantian deontological injunction to "treat humanity, whether in your own person or in that of any other, always as an end and never as a means only."[4] Crucially, this injunction says that it *is* morally permissible to treat people as means, but not *only* as means. I will go on to suggest that, if the aim of' cloning is to promote health by preventing or even curing disease in people, then the strongest arguments for cloning will pertain to (5) and (6), which effectively sidestep any ethical concerns, because the cells that are cloned and their products are nonessential parts of human beings without any moral status of their own.

II. Biological Factors

Very briefly, cloning consists in transplanting or injecting the nucleus of an adult body cell into an enucleated egg cell (oocyte) of another animal.[5] Convinced that past efforts to clone mammals had failed because the donor cell that supplied the nucleus and the recipient egg cell were at different stages of the cell cycle, Wilmut and his colleagues cut the supply of nutrients and thereby induced the donor cell to go into a "quiescent" phase in which the cell stops dividing. In this way, they were able to prepare the donor nucleus so that it would be compatible with the egg cytoplasm. This also enabled the transmuted nuclei to become reprogrammed to create every other kind of cell and remain totipotent, that is, capable of retaining all the genetic material necessary to produce a complete organism, given that every cell carries a complete set of genes for the entire organism. The reprogramming that takes place in the two-way transfer of proteins between the nucleus and the cytoplasm effects the dedifferentiation of cells and thus allows them to become reprogrammed and totipotent.

The difficulty with which Wilmut and his colleagues produced one healthy lamb is illustrated by the fact that they obtained this result after beginning their experiment with 434 sheep oocytes. Of these, only 277 adult nuclei were successfully transferred to enucleated oocytes. The success rate of one out of 434 oocytes indicates that cloning is not only an inefficient but also an exceedingly complicated biological process involving many different causal factors. In the case of human cloning, hundreds of egg cells would be required to produce one cloned individual, and this number is more than what a woman is physically capable of donating. Given the enormous failure rate, attempting to clone particular human beings would involve hundreds of failed attempts and an unknown number of viable, but deformed or damaged, offspring. Hence there are good reasons to be wary of human cloning even as a remote biological possibility. More specifically, there are at least two biological considerations that tell against the procedure. And these considerations have ethical implications insofar as they pose a risk of harm to people cloned from donor cells.

The first consideration pertains to the genetic age of the DNA of the donated nucleus that is injected into the enucleated oocyte. According to one theory of cell aging, cells follow a preset genetic recipe and divide only a certain number of times in a life.[6] This limit on cell replication is a property of the nucleus and its DNA, rather than of the cytoplasm of the cell. The aging of cells is controlled by proteins called telomeres, which are located at the ends of chromosomes and which become shorter and shorter until they no longer can protect vital parts of the chromosome. At this point, the cell stops dividing and dies. Alternatively, there is a theory which says that cells age through random change to DNA as it interacts with chemicals inside and outside the body.[7] These include chemicals in the environment, such as cigarette smoke, ionizing radiation, and pesticides, as well as chemicals inside the cell, such as hydroxyl radicals, which are waste products of cells during metabolism. These can cause mutations in the sequences of base pairs that constitute the DNA code. As a person ages, these mutations adversely affect the normal DNA repair mechanisms in cells that ordinarily correct for such damage. On this view of aging, irreversible DNA damage accumulates as the organism ages. Consequently, proteins that control the normal functioning of cells are altered, and with this there is increased likelihood of disease and premature aging.

If the first account of cell aging is correct, then a cloned individual would most likely develop normally and not age prematurely or suffer from the health effects of genetic mutation. Assuming that there is compatibility between the donor egg cytoplasm and the recipient nucleus and that the donor egg cytoplasm has been making telomerase all along, the normal activity of telomerase and therefore normal cell division and aging would obtain in the cloned individual organism. But if the donor cell nuclear DNA has been exposed to radiation, chemical mutagens, and carcinogens, then the clone's cells would be considerably older than its chronological age and it would age considerably faster than the normal rate. Such an individual would be at risk of inheriting genetic defects caused by cumulative damage in the DNA and its nuclear donor cell and consequently would be more likely to experience disease and premature aging. Significantly, these biological factors have ethical implications for the cloning of human beings. Cloning may result in the existence of a diseased or disabled individual, due to genetic mutations in the donor cell DNA, which would harm that individual by defeating his or her interest in being brought into existence in a healthy state and living out a normal lifespan without disease and disability.[8] The possible deleterious biological effects of cloning on the aging process could indirectly cause harm in a person by producing an adverse bodily condition that he would have to experience. And this could be one ethical ground on which cloning humans might be impermissible.

In addition, mutations may occur while somatic (body) cells are growing in culture. Even here, exposure to certain environmental factors of the sort I have mentioned may cause further genetic damage as well. In this scenario, it would be extremely difficult to determine whether the donor cell nucleus was normal or abnormal, no way of ascertaining the extent of damage to its DNA due to factors inside or outside the cell. There would be no way of knowing whether the cloned individual would develop normally or have to experience disease or premature aging because of genetic damage passed on to it from its donor, which would defeat its interest, once it exists, in developing normally and thus harming it. The upshot is that the random damage theory of cell aging provides rather compelling reasons against human cloning, on both biological and ethical grounds. The issue here is not whether a donor cell nucleus is a potential person with a right to be brought into existence. Rather, the issue is that a cloned individual, given that he already exists or will exist, has an interest in living out a lifespan in a normal, disease-free, way. And given that harm consists in the defeat of one's interests, a cloned individual may be harmed by having to live with a disease resulting from genetic damage to the adult donor cell that is cloned. Insofar as we can prevent this from occurring by not undertaking the cloning procedure, there may very well be a moral requirement *not* to clone. Since there is no way of knowing whether or not there is genetic damage, that merely possible people have no right to be brought into existence, and that the potential risks seem to outweigh the potential benefits, arguably we should err on the side of caution and prevent this state of affairs from obtaining by not cloning any human being into existence.

As an asexual form of reproduction, cloning also might result in serious long-term genetic costs to the survival capacity of the human species. Sexual reproduction allows for genetic variation among offspring, which in turn enhances the ability of a species to adapt to changes in the physical environment.[9] In particular, genetic variation may help organisms evolve in such a way as to cope better with infectious agents. This confers a crucial survival advantage on it species, which is then able to transfer its genes to future generations. Sexual reproduction also confers an evolutionary advantage on a species by enhancing DNA repair mechanisms in organisms belonging to that species. This results from a process known as "outcrossing," which involves the fusion of two cells in sexual reproduction so that the genetic material of each parent cell becomes enclosed inside a single membrane. Enzymes are then produced which can check for and repair damage to sequences of chromosomal base pairs. In this way, the outcrossing resulting from sexual reproduction can prevent the occurrence and expression of genetic mutations. Over many generations, cloning, as an asexual form of reproduction, may result in the expression of mutations and a higher incidence of disease and premature aging and death in human beings. More generally, it may threaten the natural variation necessary for the human species as a whole to survive in the environment and to pass on its genes to future generations. Hence evolutionary biology provides reasons against cloning.

In this section, I have addressed some of the biological risks entailed by human cloning, as well as the ethical implications of these risks for humans. This is not to suggest that all forms of cloning should be banned. On the contrary, there are compelling medical reasons for certain forms of cloning which may be of tremendous benefit to humans. I now will examine these forms and the effects they might have on persons, paying particular attention to the question of whether they are morally permissible and on what grounds they can be morally justified.

Specifically, I will consider whether they are consistent with, or violate, the Kantian injunction to treat persons not merely as means but also at the same time as ends in themselves.

III. MEANS AND ENDS

In the Introduction, I cited six possible forms of cloning, though only three of these pertained to full-bodied human organisms. The others pertained to embryos, tissues, organs, and other body parts. Let us examine each of these two types in turn.

Suppose that the parents of a deceased or dying child want to clone an individual who is genetically identical with the deceased and thus "replace" it to compensate for their loss or else carry on the family line. If the sole intent of the parents is replacement or compensation, then clearly the cloned individual would be treated as a mere means, a blatant violation of the Kantian injunction expressed in the second formulation of the Categorical Imperative. A human being is not a mere extension of its parents or siblings, and to treat him or her as such is to violate the intrinsic dignity and worth that they possess in virtue of the fact that they are human agents with the capacity for reason.[10] This would be the highest form of grandiloquent egoism and narcissism on the parents' part. Similar remarks would apply to parents who want to clone only to have a child with particular physical or cognitive traits. Yet despite the apparent repugnance of being reproduced to fill a void left by the deceased sibling, if the cloned replacement child at the same time was loved and treated with the dignity and respect commanded by its intrinsic worth, then such a procedure might be morally permissible and justifiable on Kantian grounds. Nevertheless, one might have good reason to be skeptical of the motivation for intentionally creating an individual with the same physical traits as the deceased child, leading one to question whether in fact the younger child was being treated as a mere means.

In a similar case, parents may decide to clone a brother for a son dying from a blood disorder like leukemia. Because the cloned brother would be genetically identical with the diseased sibling, the former's bone marrow would be compatible with the latter's and thus there would be no rejection by his immune system. As soon as the cloned brother's bone marrow was mature enough for harvesting, it could be extracted and transferred to the older brother to cure him and thereby save his life. This recalls the case of Anissa Ayala of Walnut Grove, California in the late 1980s.[11] While Anissa was slowly dying from leukemia, her parents decided to have another child (a daughter) on the slight chance that it would be a donor match with Anissa. Fortunately, the younger daughter's bone marrow was compatible with Anissa's, the marrow was transferred, and Anissa's life was saved. While at first blush the parents' decision and action seemed morally repugnant and unjustifiable, all accounts now suggest that the child who was conceived in order to save her sibling's life has been loved and respected by her parents and sister as a distinct individual with intrinsic worth of her own. In other words, the Ayalas *were* treating the younger daughter as a means. Yet they did not treat her *merely* as a means but also as an end in itself.

Similarly, in the other case that I have mentioned, simply cloning a genetically identical but numerically distinct individual to save or replace another by itself does not necessarily imply that that individual is treated solely as a means. If he or she is also treated as an end in itself, then the procedure of cloning a human being is not in itself morally objectionable. For on the Kantian view, treating an individual as an end in itself implies respect for that person as an autonomous locus of dignity and intrinsic worth, one whose capacity for reason enables it to have interests and rights of its own. If respecting a person means upholding its interests and rights, and if cloning does not thwart its interests or violate its rights, then a person is not harmed or wronged just because it is cloned.

Lesbian and gay couples might also decide to have a cloned child so that it would be biologically connected to at least one of them. One partner would provide the nucleus, the other the mitochondrial DNA and cytoplasm. Here, though, there would have to be compelling reasons for the intrinsic value of genetic relatedness between parent and child to justify producing a cloned child. Artificial insemination or in vitro fertilization may produce the same result, though without the same degree of biological relatedness entailed by cloning. Another scenario in which cloning might be medically and morally justified is one in which a couple is infertile, where the male has immature sperm or an extremely low sperm count, or else the female's fallopian tubes are blocked. Even here, though, cloning would be justified only if the couple had exhausted all other forms of assisted reproduction and that they do not undergo the procedure *solely* to have a child with particular traits.

Creating a genetically identical clone with the same physical traits as its parents or a sibling is not by itself inherently unethical because genetic identity is not equivalent to personal identity and thus does not threaten the distinctiveness of persons. Having the same DNA as a parent donor may give a cloned child the same physical features as that parent. But personal identity consists in the persistence through time of one and the same individual identified with a body, brain, and a set of mental states including desires, beliefs, intentions, and memories unified from the standpoint of the conscious present.[12] The nature and content of these mental states are shaped as much by, if not more so than, one's social and physical environment as they are by one's genetic makeup. The genes that a cloned individual inherits from a parent or sibling may *influence* the psychological properties in which personal identity consists by shaping the physical properties of the body and brain that support one's psychology. But genes alone cannot account for the uniqueness of one's cognitive and emotional experience of and response to the environment and therefore do not completely *determine* the nature and content of the mental states that constitute one's personal identity. To be sure, genes only have their effect on personhood and personal identity within a particular environment. Nevertheless, it is unlikely that genes could be the primary explanation for one's psychological properties owing to the complexity in the way that environmental actors play a causal role in the etiology of one's desires, intentions, and memories. Nor can genes account for the phenomenological quality of being conscious of having these mental states. In addition, the fact that parent and child will be of different generations and thus develop their psychology in different environments means that their mental states will be qualitatively distinct.

At least one empirical study of the psychological traits of genetically identical twins supports these claims. In a questionnaire devised by psychologist Thomas Bouchard, there was a 50 percent correlation between the personality traits of genetically identical twins, based on different responses to the same questions.[13] This suggests that at least half of one's psychological traits are due to environmental factors independently of or in conjunction with genes. Genetic identity alone underdetermines the sense of a self which develops through the construction of a unified life plan with goals and experiences that confer meaning on one's life. In Bouchard's words, "Selves, unlike cells, can never be cloned." Ultimately, personhood and personal identity are not solely functions of genetic ancestry.

One of the main concerns about cloning human beings is that it threatens to undermine a child's freedom, its autonomy in constructing and carrying out a life plan. But however egocentric or otherwise misguided parents' aims may be in wanting to produce a clone of themselves, genetic identity alone would not necessarily restrict or undermine the child's freedom to develop as an inviolable self. What restricts or undermines a child's freedom, rather, is the parents' refusal or inability to allow the child to develop his or her own plans and values by imposing their own plans and values on the child in an attempt to control its life. This phenomenon occurs all too often in many families. But surely it is not confined to a cloning scenario, because it is not simply a function of the biological relatedness between parent and child, or of the child's genetic identity.

Only if the production of genetic identity between parent and child through cloning were an integral part of the parent's plan to control the child's life would cloning and genetic identity be morally objectionable, on the ground that they would violate the child's autonomy. By itself, though, genetic identity is morally neutral, no moral status attaches to it. After all, no moral judgments attach to the fact that genetically identical twins result from the monozygotic twinning of an embryo in the natural, unassisted, form of human reproduction. This is largely because the twins go on to develop as numerically and qualitatively distinct individuals in virtue of developing distinct psychological properties and selves over time. And yet in the case of genetically identical twins, one is effectively a clone of the other. Still, it is important to distinguish *natural* from *deliberate* production of offspring. The latter has moral status insofar as it presupposes an intention on the part of a parent, parents, or scientist to engineer a being with particular traits for a purpose that conforms to certain values or ideals. The content of the intention to deliberately produce a genetically identical copy of an existing individual may be either beneficial (as in the Ayala case) or harmful (as in the narcissistic parent's desire to create a being in its own image). As a deliberate form of reproduction, cloning therefore *does* have moral status. But there is nothing inherently immoral or morally objectionable about cloning itself. If it were to be banned on moral grounds, then it would only be because cloned offspring would be produced and treated as mere means to one of the morally objectionable ends that I have cited. Or, as I reasoned in section II, it could be banned because of the likely genetic damage to cells which subsequently resulted in harm to the individuals who would be cloned.

It is worth emphasizing that the National Bioethics Advisory Committee, while recommending a ban on cloning human beings, did not also recommend that research be restricted on the cloning of human cells for the purpose of producing tissues, organs, and bones. This is significant because by cloning these parts of humans, which are not essentially identical with humans and have no intrinsic moral status of their own, we effectively sidestep any disturbing ethical implications of cloning full-bodied human beings. With the possible exception of couples who cannot conceive a child through any other means of assisted reproduction, the most salient reason for cloning is for therapeutic purposes. That is, the point of cloning cells from already existing people for the purpose of creating tissues, bones, or organs would be to cure these same people of disease due to failing organs, tissues, or genetically defective cells, and in some cases to prevent them from becoming diseased in the first place. If this practice were to become feasible, then there would be no compelling biological or medical need to clone human beings. The purpose of cloning would not be to make more people, but rather to make more already existing people healthy.[15]

It would be difficult to find anything morally objectionable about cloning human cells for the therapeutic purpose of curing or preventing disease. Such a procedure could be morally objectionable for either of two reasons: (1) either the procedure would result in some form of harm to existing people; (2) or it somehow would undermine or violate their essential humanity. With respect to the first point, if the procedure in question is by definition therapeutic and can cure or prevent disease with no deleterious side-effects, and if people have an interest in avoiding disease, then the procedure itself would not be harmful to people. With respect to the second point, human body parts are not identical with persons; personal identity does not consist of cells and other body parts. What makes us humans, or individual persons who persist through time, are the psychological properties associated with our desires, beliefs, intentions, emotions, and memories. And while these psychological properties causally depend on biological body parts like cells, they are not reducible to them. Put another way, persons *have* body parts; but they *are* not body parts and cannot be described entirely in bodily terms. To be sure, humans are embodied beings and in this respect our humanity causally depends on our biology. Nevertheless, it is in virtue of our psychology that we are rational and moral agents who can construct selves that persist through time. To the extent that cells and other body parts that we have but are not identical with us and have no moral status of their own can be manipulated in a cloning procedure, doing so does not violate or undermine our essential humanity, which is a function of our psychology. Furthermore, at the biological level of cells, tissues, bones, and organs, there are no individuals with rights or interests who could be violated, harmed, or wronged. Still, it is because of the causal dependence of our psychology on our biology that cloning these body parts can benefit humans, given our interest in not having to suffer from disease, disability, or premature aging and death.

In theory, medical researchers in the future might be able to remove a small amount of a particular tissue from a patient, dedifferentiate the cells in culture, and then genetically reprogram them to differentiate into a specific kind of cell for a particular organ which could be recognized by the immune system and not be rejected by it. For instance, it may be possible for an only child with polycystic kidney disease, whose parents are unable to donate their organs to him, to have cultured kidney cells from his organs' tissue genetically reprogrammed to replace the genetically defective cells causing the disease. This would avoid having to clone another human being who may or may not be a compatible kidney donor. In an analogue of the Ayala case, genetically defective stem cells causing the leukemia could be extracted and genetically reprogrammed in culture, after which they could be injected back into the afflicted individual and thereafter produce normal amounts of white blood cells. This would preempt the need to bring another human being into existence for a therapeutic purpose and would avoid the ethical question of whether we would be treating such a person as a mere means. And because the procedure would be an autologous transplant, it would preclude the threat of rejection by the immune system. Still another possibility of the therapeutic use of cloning cells to produce tissues and organs might involve defective livers. The technique of autologous hepatocyte transplantation uses the patient's own liver cells. Cells from portions of tissue from a liver diseased by defective genes could be removed from the tissue and genetically modified in culture to repair the defective gene. They then could be infused back into the patient's liver, merging with the organ and replacing the defective hepatocytes.[16]

If patients suffering from kidney disease, leukemia, or liver failure were too sick to benefit from such a procedure, or if the genetic damage in cells were such that the cells could not be reprogrammed, then at an earlier time before the onset of symptoms they could store cells for future production of replacement tissues or organs. Of course, this would assume that they knew they were likely to contract a disease given the genetic information available to them. This is not to deny that we should proceed cautiously on research in this area, given the probability of DNA damage to cells from storing them for long periods of time. Nevertheless, in an era when large numbers of people die every day for need of organ transplants, cloning organs would provide a medical opportunity for meeting this need.

Recently, molecular biologists led by Jonathan Slack at Bath University created a frog embryo without a head.[17] They achieved this result by manipulating certain genes in such a way as to suppress development of a tadpole's head, trunk, and tail. This leads one to speculate whether the technique could be adapted to grow human organs such as hearts, kidneys, and livers in an embryonic sac living in an artificial womb. If such a technique were to become feasible, then people needing organ transplants could have organs grown from their own cloned cells. The embryo could be genetically reprogrammed to suppress all the parts of the body except for the needed parts, plus a heart and blood circulation. This could ameliorate the problem of the shortage of organs for transplantation. Even if one were to argue that embryos have moral status and should not be interfered with in any way, growing partial embryos could bypass legal restrictions and ethical concerns. For without a brain or central nervous system, these organs would not technically qualify as embryos. Insofar as human body parts like cells and tissues lack moral status, manipulating them is not morally objectionable. This type of genetic manipulation would be morally objectionable if it resulted in harmful effects on persons in the form of disease or disability over the course of their lives.

Furthermore, the cloning of cells could be useful in correcting mutations for genetic diseases in early embryonic tissue growing in culture. Take sickle-cell anemia (SCA), for example. At the embryonic stage of a developing human organism, when genetic testing has shown that cell nuclei have the genetic abnormality causing the disease, a normal functioning gene for the blood's oxygen-carrying protein, beta globin, which is mutated in SCA, could be inserted into embryonic cells by means of a viral vector. The DNA of one of the cell nuclei could then be implanted into a new enucleated egg cell from the mother and a new pregnancy could begin. In this way, the original embryo would be replaced by a genetically healthy clone of itself, which could then develop into a healthy human being.[18] Although the new embryo would be genetically identical with the original, the individual developing from the genetically normal embryos would likely go on to develop a different set of psychological properties and thus would be a qualitatively and numerically distinct person. Still, what is at issue here is the motivation for the procedure, which is to prevent disease in people by intervening genetically before any person comes into existence. The aim is to prevent harmful effects on people by eliminating the genetic cause at a very early stage of embryonic development. And it is both medically and morally preferable to prevent disease from occurring in the first place than to bring people into existence and then try to treat them for or cure them of a disease. In this way, we prevent people from being harmed at all.

IV. A Slippery Slope? Negative and Positive Eugenics

Thus far, I have argued that while cloning cells to produce full-bodied human beings is morally permissible only in rare cases and only on the condition that cloned individuals are not used merely as means, cloning of cells to produce tissues, bones, or organs may be morally permissible in many cases. The underlying rationale for this claim is that we should use cloning, not to create more people, but to make more actually existing people healthy, either by preventing disease or by curing it. Even here, though, one might ask whether the cloning of tissues and organs would be motivated by the desire to improve the human species and thus serve eugenic goals. If so, then would there not be moral grounds for banning the practice, even for the production of body parts?

It is important to distinguish positive from negative eugenics. The aim of negative eugenics is to promote health by preventing disease through genetic or other biological forms of intervention in the life of a human organism. In this sense, it is motivated by beneficence. Gene therapy to deliver the factor necessary for normal blood clotting in hemophiliacs is an example of the first type of intervention. Genetic reprogramming of cultured kidney cells in a person suffering from polycystic kidney disease would be an example of the second type. By contrast, the aim of positive eugenics is to enhance or raise the normal level of cognitive and physical functioning for the human species. This likely would come through genetic manipulation of cells to create exceptional physical or cognitive traits. In this sense, it is motivated by perfectionism.

Some might claim that there is a gray area between disease and health where our intuitions are not all that clear. This makes it difficult to differentiate between health promotion and enhancement and, in turn, between negative and positive eugenics. Accordingly, definitions of "health" and "disease" are in order. Perhaps the most plausible definition of health is in terms of cognitive and physical functioning which give one opportunities for formulating a life plan, undertaking and completing projects within that plan, and thereby achieving a decent minimum level of well-being.[19] Conversely, disease is defined in terms of cognitive or physical dysfunction that significantly limits one's opportunities for achieving the decent minimum. For example, someone with mild seasonal mood swings might be slightly depressed but still able to form and execute intentions in actions which enable him to realize opportunities for achievement. Such a person would be considered healthy in cognitive terms. On the other hand, someone diagnosed with severe clinical depression and unable to form and execute intentions in actions would not be able to realize such opportunities and thus would be considered diseased in cognitive terms. In the first case, the person's level of cognitive functioning would be normal, and efforts to raise this level would amount to cognitive enhancement. In the second case, the person's level of cognitive functioning would not be normal and efforts to raise it to the normal level would amount to cognitive health promotion.

Assuming that we are motivated by beneficence rather than perfectionism in considering people's health, there are moral reasons to ensure that people have adequate functioning for opportunities that will enable them to achieve a decent minimum level of well-being. These reasons are stronger the lower people fall below that level, and weaker the higher they fall above this level. Consistent with beneficence, we are not morally required to raise people's cognitive and physical functioning *above* the level of a decent minimum, but only *up to* that level.

In addition to raising questions about the implications of enhancement for what it means to be human, a perfectionist program of positive eugenics would be morally objectionable if it implied that the sickest and most vulnerable groups (people with physical disabilities, the mentally retarded, the poorest of the poor) in our species would be disvalued on the ground that their physical and cognitive level of functioning would only serve to lower the average level of functioning for the species as a whole. This could lead to policies that might violate their intrinsic worth as humans, as well as their right to have lives that, for many of them, are very much worth living. Indeed, this scenario may provide grounds for a moral prohibition against positive eugenics. Insofar as we are motivated by beneficence rather than perfectionism, it is morally permissible, perhaps even obligatory, to practice negative eugenics by creating people *without* particular traits, or by *removing* these traits once they exist if these traits cause disease or disability. By doing so, we enable them to have the cognitive and physical functioning necessary to achieve a decent minimum level of well-being. On the other hand, it is not only not morally obligatory but indeed impermissible to create people *with* particular traits, or to manipulate traits in existing people, if the sole reason for doing so is to raise the normal level of human species functioning by creating exceptional cognitive or physical traits in people.[20]

Reflecting on the eugenic aspects of cloning, Leon Kass says that "we do indeed already practice negative eugenic selection, through genetic screening and prenatal diagnosis. Yet our practices are governed by a norm of health."[21] Kass goes on to draw the crucial distinction between health promotion and genetic enhancement, suggesting that while the latter is morally repugnant and unjustifiable, the former is not only morally justifiable but obligatory. But how can we be so sure that a program of negative eugenics will not evolve into a program of positive eugenics? Is there not a slippery slope here?

Bernard Williams points out that the slippery slope argument "is often applied to matters of medical practice. If X is allowed, the argument goes, then there will be a natural progression to Y."[22] For present purposes, we can take X to represent negative eugenics and Y to represent positive eugenics. This leads to the "horrible result" of what we find at the bottom of the slope, which would imply a repeat of the inhumane treatment of people and the violation of their intrinsic worth which have occurred in recent history, with the ostensible aim of enhancing people's normal cognitive and physical functioning. Presumably, what makes any eugenic slope slippery is that once we get on the negative side, we cannot get off and inevitably fall to the positive side. The point of the argument is that we should not start to begin with, that negative eugenics is not morally justifiable and accordingly should not be practiced, since it inevitably leads to positive eugenics.

But we can make a reasonable and clear distinction between preventing diseases and thereby promoting health to a decent minimum, on the one hand, and enhancing people's cognitive and physical traits and thereby raising the level of normal functioning, on the other. These are distinct aims generated by completely different sets of reasons corresponding to beneficence and perfectionism, respectively. Provided that the rationale for cloning is clearly spelled out in terms of promoting people's health so that they can achieve a decent minimum level of health and well-being, and that public policy guidelines and laws are carefully formulated

and implemented to restrict cloning to disease prevention, cure, and health promotion, there is no reason to believe that cloning will inevitably lead to morally
repugnant forms of enhancement. Moreover, provided that these guidelines and
laws protect the sick and disabled, the most vulnerable groups in society, with equal
access to the different forms of cloning, and where there is no socially coercive
pressure to create or improve existing people according to perfectionist ideals,
cloning as a form of negative eugenics would be a morally justifiable program.[23]

V. CONCLUSION

The biological possibility of cloning human cells to produce tissues, bones,
organs, and even humans who are genetically identical with their donors offers
tremendous opportunities for medicine, especially with respect to preventing,
treating, or even curing genetically caused diseases. But even if it is at most a
remote possibility, the ethical implications of cloning complete human organisms
make us well-advised to proceed with caution on this front. For if we clone individuals solely for the purpose of making genetic and physical duplicates of ourselves in our own image, or to replace a deceased or dying person, then the practice would be morally repugnant and as such impermissible on the ground that we
would be using the cloned individual as a mere means and not also as an end in
itself. This would violate the Kantian injunction that has been invoked as a guiding
ethical principle in these matters. It is in this respect that cloning would threaten
to undermine our essential humanity.

To the extent that cloning is motivated by therapeutic goals, and that cells can
be reprogrammed and cloned to produce genetically identical copies of tissues or
organs, the cases in which there would be compelling medical reasons to clone full-
bodied human beings would be very rare. This effectively would sidestep the most
troublesome ethical questions generated by cloning. Significantly, cloning cells to
produce these body parts would not be inherently unethical because body parts have
no ethical status of their own. Nor would such a procedure threaten our essential
humanity because persons have but are not reducible to and thus not identical with
body parts. Genetic identity is not personal identity; selves cannot be cloned.
Although the psychological properties associated with the emotions, beliefs, desires,
intentions, and memories which constitute personal identity over time causally
depend on biological properties involving the body and brain, our psychology
cannot be completely described or explained in biological terms. Our personhood,
our humanity, is not merely a function of the body's tissues and organs.

CLONING

The only respects in which cloning on this level might be morally objectionable would be if the process entailed DNA damage to cell nuclei which manifested itself in premature aging and disease, or if this form of asexual reproduction entailed long-term deleterious effects on the survival capacity of the human species. Even here, though, the impermissibility would pertain not to the way in which tissues and organs are produced, but instead to the potential harmful effects it might have on existing persons. Provided that it is consistent with a negative eugenics program of health promotion through disease prevention, cloning can be a morally justifiable practice that accords with our deepest ethical convictions about persons and the value of their lives.[24]

NOTES

1. Ian Wilmut et al., "Viable Offspring Derived from Fetal and Adult Mammalian Cells," *Nature* 385 (February 27, 1997): 310–13. See also the accompanying article by Colin Stewart, "An Udder Way of Making Lambs," ibid., pp. 769–71.

2. James Childress, Susan M. Wolf, Courtney S. Campbell, Daniel Callahan, and Erik Parens discuss different aspects of the report in "Cloning Human Beings: Responding to the National Bioethics Advisory Commission's Report," *Hastings Center Report* 27 (September–October 1997): 9–22.

3. Perhaps the most outspoken critic of human cloning who argues along these lines is Leon Kass, "The Wisdom of Repugnance: Why We Should Ban the Cloning of Humans," *New Republic* (June 2, 1997): 17–26.

4. This is the second formulation of the Categorical Imperative. The first formulation says: "I ought never to act in such a way that I could not also will that my maxim should be a universal law." From the *Foundations of the Metaphysics of Morals* (1785), trans. L. W. Beck, 2d ed. (New York: Macmillan, 1990), p. 429, for the second formulation of the CI, p. 402 for the first. Page references are to the Royal Prussian Academy Edition. F. M. Kamm defends the Kantian notion of a person as an inviolable end-in-itself at considerable length in *Morality, Mortality*, vol. 2: *Rights, Duties, and Status* (Oxford: Oxford University Press, 1996), and "Nonconsequentialism, the Person as an End-in-Itself, and the Significance of Status," *Philosophy and Public Affairs* 21 (1992): 354–89.

5. See Wilmut et al., "Viable Offspring Derived from Fetal and Adult Mammalian Cells." Also, Marie A. Di Berardino and Robert G. McKinnell, "Backward Compatible," in *The Sciences* 37, no. 5, Special Issue: "The Promise and Peril of Cloning" (September–October 1997): 32–37, at p. 33.

6. This is what is known as the "Hayflick Limit," named after the biologist, Leonard Hayflick, who proposed and defended the theory of programmed cell death. See his "The Cellular Basis for Biological Aging," in Leonard Hayflick and C. E. Finch, eds., *Handbook of*

the Biology of Aging (New York: Van Nostrand, 1977), pp. 159–86. Ronald Hart, Angelo Turturro, and Julian Leakey discuss the genetic aspects of aging in "Born Again?" in *Promise and Peril,* pp. 47–51. See also D. Broccoli and H. Cooke, "Aging, Healing, and the Metabolism of Telomeres," *American Journal of Human Genetics* 52 (1993): 657–60; and B. W. Stewart, "Mechanisms of Apoptosis: Integration of Genetic, Biochemical, and Cellular Indicators," *Journal of the National Cancer Institute* 86 (1994): 1286–96. John Medina offers a more general discussion of the biology of aging in *The Clock of Ages* (Cambridge: Cambridge University Press, 1996).

7. Hart, Turturro, and Leakey, ibid.

8. In *Harm to Others* (Oxford: Oxford University Press, 1984), pp. 79–95, Joel Feinberg discusses the notion of harm as the defeat of one's interests.

9. See Richard E. Michod, "What Good Is Sex?" in *Promise and Peril*, pp. 42–46. Compare Michod's discussion with the very different interpretation of the value of sexuality given by Kass in "The Wisdom of Repugnance."

10. Kant explains intrinsic worth in these terms in the *Foundations*, pp. 394 ff.

11. Cited by Philip Kitcher, "Whose Self Is It, Anyway?" in *Promise and Peril*, pp. 58–62.

12. For representative versions of the "psychological continuity view" of personal identity, see Derek Parfit, *Reasons and Persons* (Oxford: Clarendon Press, 1984), parts II and III; Thomas Nagel, *The View from Nowhere* (New York: Oxford University Press, 1986); and Peter Unger, *Identity, Consciousness, and Value* (New York: Oxford University Press, 1990). There are differences among these versions, but I ignore them for present purposes.

13. "Genes, Environment, and Personality," *Science* 246 (June 17, 1994): 1700–1701.

14. "Whenever the Twain Shall Meet," in *Promise and Peril*, pp. 52–57, at p. 54.

15. This idea derives from Jan Narveson's claim that we do not have a moral duty to make happy people, but only to make people happy. He argued that the benefit of an act is the good it brings to already existing people and does not include the good of people who may come into existence as a result of the act. See "Utilitarianism and New Generations," *Mind* 76 (1967): 62–72, and "Moral Problems of Population," *Monist* 57 (1973): 62–86. David Heyd offers a more recent analysis of these matters in *Genethics: Moral Issues in the Creation of People* (Berkeley: University of California Press, 1992).

16. Hart, Turturro, and Leakey, "Born Again?" p. 51.

17. "The Cloning of a Headless Frog," *London Sunday Times*, October 19, 1997.

18. See Steve Mirsky and John Rennie, "What Cloning Means for Gene Therapy," *Scientific American* (June 1997): 122–23.

19. The account of health in terms of functioning which I use derives from the empirical (nonnormative) accounts of Christopher Boorse, "Health as a Theoretical Concept," *Philosophy of Science* 44 (1977): 542–71; and Leon Kass, *Toward a More Natural Science* (New York: Free Press, 1985). A much broader (too broad, I believe) definition of health is given in the Preamble to the Constitution of the World Health Organization (1986): "Health is a state of complete physical, mental, and social well-being and not merely the absence of disease or infirmity."

20. For general discussion of eugenics, see, for example, Jonathan Glover, *What Sort of People Should There Be?* (Harmondsworth: Penguin, 1984); Philip Kitcher, *The Lives to Come: The Genetic Revolution and Human Possibilities* (New York: Simon and Schuster, 1996); and Leroy Walters and Julie Gage Palmer, *The Ethics of Human Gene Therapy* (New York: Oxford University Press, 1997), ch. 4. Daniel Kevles examines the history of eugenics in the United States in *In the Name of Eugenics: Genetics and the Uses of Human Heredity* (New York: A. A. Knopf, 1985).

21. Kass, "The Wisdom of Repugnance," p. 24.

22. "Which Slopes Are Slippery?" in *Making Sense of Humanity* (Cambridge: Cambridge University Press, 1995), pp. 213–23, at p. 213. See also Frederick Schauer, "Slippery Slopes," *Harvard Law Review* 99 (1985), pp. 361–83; and Wibren van den Burg, "The Slippery Slope Argument," *Ethics* 102 (October 1991): 42–65.

23. My views are similar to what Kitcher calls "utopian eugenics" in *The Lives to Come*, pp. 201–203.

24. I am grateful to Stephen Clark for very helpful comments on an earlier draft.

BIBLIOGRAPHY

Boorse, C. 1997. "Health as a Theoretical Concept." *Philosophy of Science* 44: 542–71.
Bouchard, T. 1997. "Whenever the Twain Shall Meet." *The Sciences* 37 (September–October), Special Issue, *The Promise and Peril of Cloning*, 52–57.
———. 1994. "Genes, Environment, and Personality." *Science* 246 (June 17): 1700–1701.
Broccoli, D., and H. Cooke. 1993. "Aging, Healing, and the Metabolism of Telomeres." *American Journal of Human Genetics* 52: 657–60.
Childress, J., et al. 1997. "Cloning Human Beings: Responding to the National Bioethics Advisory Commission's Report." *Hastings Center Report* 27 (September–October): 9–22.
Cloning Human Beings: The Report and Recommendations of the National Bioethics Advisory Commission. Rockland, Md., June 1997.
Di Berardino, M., and R. McKinnell. 1997. "Backward Compatible." *The Sciences* 37: 32–37.
Feinberg, J. 1984. *Harm to Others.* Oxford: Oxford University Press.
Glover, J. 1984. *What Sort of People Should There Be?* Harmondsworth: Penguin.
Hart, R., A. Turturro, and Julian Leakey. 1997. "Born Again." *The Sciences* 37: 47–51.
Hayflick, L. 1977. "The Cellular Basis for Biological Aging." In L. Hayflick and C. Finch, eds. *Handbook of the Biology of Aging*, pp. 159–86. New York: Van Nostrand.
Heyd, D. 1992. *Genethics: Moral Issues in the Creation of People.* Berkeley: University of California Press.
Kant, I. 1785/1990. *Foundations of the Metaphysics of Morals*, 2d ed., trans. L. W. Beck. New York: Macmillan.
Kass, L. 1997. "The Wisdom of Repugnance: Why We Should Ban the Cloning of Humans." *New Republic* (June 2): 17–26.
———. 1985. *Toward a More Natural Science.* New York: Free Press.

Kevles, D. 1985. *In the Name of Eugenics: Genetics and the Uses of Human Heredity.* New York: A. A. Knopf.

Kitcher, P. 1997. "Whose Self Is It, Anyway?" *The Sciences* 37: 58–62.

———. 1996. *The Lives to Come: The Genetic Revolution and Human Possibilities.* New York: Simon and Schuster.

Medina, J. 1996. *The Clock of Ages.* Cambridge: Cambridge University Press.

Michod, R. 1997. "What Good Is Sex?" *The Sciences* 37: 42–46.

Mirsky, S., and J. Rennie. 1997. "What Cloning Means for Gene Therapy." *Scientific American* (June): 122–23.

Nagel, T. 1986. *The View From Nowhere.* New York: Oxford University Press.

Narveson J. 1967. "Utilitarianism and New Generations." *Mind* 76: 62–72.

———. 1994. 1973. "Moral Problems of Population." *Monist* 57: 62–86.

Parfit, D. 1984. *Reasons and Persons.* Oxford: Clarendon Press.

Stewart, B. 1994. "Mechanisms of Apoptosis: Integration of Genetic, Biochemical, and Cellular Indicators." *Journal of the National Cancer Institute* 86: 1286–96.

Stewart, C. 1997. "An Udder Way of Making Lambs." *Nature* 385 (February): 769–71.

Unger, P. 1990. *Identity, Consciousness, and Value.* New York: Oxford University Press.

Walters, L., and J. Palmer. 1997. *The Ethics of Human Gene Therapy.* New York: Oxford University Press.

Williams, B. 1995. "Which Slopes Are Slippery?" In *Making Sense of Humanity,* pp. 213–23. Cambridge: Cambridge University Press.

Wilmut, I., et al. 1997. "Viable Offspring Derived From Fetal and Adult Mammalian Cells." *Nature* 385 (February): 310–13.

14 Cloning and Human Dignity
John Harris

T he panic occasioned by the birth of Dolly sent international and national bodies and their representatives scurrying for principles with which to allay imagined public anxiety. It is instructive to note that principles are things of which such people and bodies so often seem to be bereft. The search for appropriate principles turned out to be difficult since so many aspects of the Dolly case were unprecedented. In the end, some fascinating examples of more or less plausible candidates for the status of moral principles were identified; central to many of them is the idea of human dignity and how it might be affected by human mitotic reproduction.[1]

Typical of appeals to human dignity was that contained in the World Health Organization statement on cloning issued on March 11, 1997:

Reprinted from *Cambridge Quarterly of Healthcare Ethics* 7 (Spring 1998): 163–67. Copyright © 1998 Cambridge University Press. Reprinted with the permission of Cambridge University Press.

WHO considers the use of cloning for the replication of human individuals to be ethically unacceptable as it would violate some of the basic principles which govern medically assisted procreation. These include respect for the dignity of the human being....

Appeals to human dignity are, of course, universally attractive; they are also comprehensively vague. A first question to ask when the idea of human dignity is invoked is: whose dignity is attacked and how? If it is the duplication of a large part of the human genome that is supposed to constitute the attack on human dignity, or where the issue of "genetic identity" is invoked, we might legitimately ask whether and how the dignity of a natural twin is threatened by the existence of her sister and what follows as to the permissibility of natural monozygotic twinning? However, the notion of human dignity is often linked to Kantian ethics and it is this link I wish to examine more closely here.

A typical example, and one that attempts to provide some basis for objections to cloning based on human dignity, was Axel Kahn's invocation of this principle in his commentary on cloning in *Nature*. Kahn, a distinguished molecular biologist, helped draft the French National Ethics Committee's report on cloning. In *Nature* Kahn states:

> The creation of human clones solely for spare cell lines would, from a philosophical point of view, be in obvious contradiction to the principle expressed by Emmanuel Kant: that of human dignity. This principle demands that an individual —and I would extend this to read human life—should never be thought of as a means, but always also as an end. Creating human life for the sole purpose of preparing therapeutic material would clearly not be for the dignity of the life created.[2]

The Kantian principle, invoked without any qualification or gloss, is seldom helpful in medical or bioscientific contexts.[3] As formulated by Kahn, for example, it would surely outlaw blood transfusions. The beneficiary of blood donation, neither knowing of, nor usually caring about, the anonymous donor uses the blood (and its donor) exclusively as a means to her own ends. The blood in the bottle has after all less identity, and is less connected with the individual from which it emanated, than the chicken "nuggets" on the supermarket shelf. An abortion performed exclusively to save the life of the mother would also, presumably, be outlawed by this principle.

CLONING

INSTRUMENTALIZATION

This idea of using individuals as a means to the purposes of others is sometimes termed "instrumentalization," particularly in the European context. The advisers to the European Commission on the ethical implications of biotechnology, for example, in their statement on ethical aspects of cloning techniques use this idea repeatedly.[4] Referring to reproductive human cloning, paragraph 2.6 states:

> Considerations of instrumentalization and eugenics render any such acts ethically unacceptable.

Applying this idea coherently or consistently is not easy! If someone wants to have children in order to continue their genetic line do they act instrumentally? Where, as is standard practice in IVF, spare embryos are created, are these embryos created instrumentally?

Kahn responded in the journal *Nature* to these objections.[5] He reminds us, rightly, that Kant's famous principle states: "respect for human dignity requires that an individual is *never* used . . . *exclusively* as a means" and suggests that I have ignored the crucial use of the term "exclusively." I did not, of course, and I'm happy with Kahn's reformulation of the principle. It is not that Kant's principle does not have powerful intuitive force, but that it is so vague and so open to selective interpretation and its scope for application is consequently so limited that its utility as one of the "fundamental principles of modern bioethical thought," as Kahn describes it, is virtual zero.

Kahn himself rightly points out that debates concerning the moral status of the human embryo are debates about whether embryos fall within the *scope* of Kant's or indeed any other moral principles concerning persons; so the principle itself is not illuminating in this context. Applied to the creation of individuals who are, or will become autonomous, it has limited application. True, the Kantian principle rules out slavery, but so do a range of other principles based on autonomy and rights. If you are interested in the ethics of creating people, then, so long as existence is in the created individual's own best interests, and the individual will have the capacity for autonomy like any other, the motives for which the individual was created are either morally irrelevant or subordinate to other moral considerations. So that even where, for example, a child is engendered exclusively to provide "a son and heir" (as is often the case in many cultures) it is unclear how or whether Kant's principle applies. Either other motives are also attributed to the parent to square parental purposes with Kant, or the child's eventual autonomy and its dear and substantial interest in or benefit from existence take precedence over the comparatively trivial issue of parental motives. Either way the "fundamental principle of modern bioethical thought" is unhelpful.

It is therefore strange that Kahn and others invoke it with such dramatic assurance or how anyone could think that it applies to the ethics of human cloning. It comes down to this: Either the ethics of human cloning turn on the creation or use of human embryos, in which case as Kahn himself says "in reality the debate is about the status of the human embryo" and Kant's principle must wait upon the outcome of that debate; or, it is about the ethics of producing clones who will become autonomous human persons. In this latter case, as David Shapiro also comments,[6] the ethics of their creation are, from a Kantian perspective, not dissimilar to other forms of assisted reproduction, or, as I have suggested, to the ethics of the conduct of parents concerned exclusively with producing an heir, or preserving their genes, or, as is sometimes alleged, making themselves eligible for public housing. Debates about whether these are *exclusive* intentions can never be definitively resolved.

Kahn then produces a bizarre twist to the argument from autonomy. He defines autonomy as "the indeterminability of the individual with respect to external human will" and identifies it as one of the components of human dignity. This is, of course, hopeless as a definition of autonomy—those in persistent vegetative state (PVS) and indeed all newborns would on such a view have to count as autonomous! However, Kahn then asserts:

> The birth of an infant by asexual reproduction would lead to a new category of people whose bodily form and genetic make-up would be exactly as decided by other humans. This would lead to the establishment of an entirely new type of relationship between the "created" and the "creator" which has obvious implications for human dignity.

Kahn is, I'm afraid, wrong on both counts. As Robert Winston has noted: "even if straight cloning techniques were used, the mother would contribute important constituents—her mitochondrial genes, intrauterine influences, and subsequent nurture."[7] These, together with the other influences, would prevent exact determination of bodily form and genetic identity. For example, differences in environment, age, and anno Domini between clone and cloned would all come into play.

Lenin's embalmed body lies in its mausoleum in Moscow. Presumably a cell of this body could be denucleated and Lenin's genome cloned. Could such a process make Lenin immortal and allow us to create someone whose bodily form and genetic makeup, not to mention his character and individuality, would be "exactly as decided by other human beings?" I hope the answer is obvious. Vladimir Ilyich Ulyanov was born on 10 April 1870 in the town of Simbirsk on the Volga. It is this person who became and who is known to most of us as V. I. Lenin. Even with this man's genome preserved intact we will never see Lenin again. So many of the things that made Vladimir Ilyich what he was cannot be reproduced, even if his genome can. We cannot recreate prerevolutionary Russia. We cannot simulate his environment and education; we cannot recreate his parents to bring him up and influence his development so profoundly as they undoubtedly did. We cannot make the thought of Karl Marx seem as hopeful as it must then have been; we cannot, in short, do anything but reproduce his genome and that could never be nearly enough. It may be that "manners maketh man" but genes most certainly do not.

As we know from monozygotic twins, autonomy is unaffected by close similarity of bodily form and matching genome. The "indeterminability of the individual with respect to external human will" will remain unaffected by cloning. Where then are the obvious implications for human dignity?

When Kahn asks: "is Harris announcing the emergence of a revisionist tendency in bioethical thinking?" the answer must be, rather, I am pleading for the emergence of "bioethical *thinking*" as opposed to the empty rhetoric of invoking resonant principles with no conceivable or coherent application to the problem at hand.

Clearly, the birth of Dolly and the possibility of human equivalents have left many people feeling not a little uneasy, if not positively queasy at the prospect. It is perhaps salutary to remember that there is no necessary connection between phenomena, attitudes, or actions that make us uneasy, or even those that disgust us, and those phenomena, attitudes, and actions that there are good reasons for judging unethical. Nor does it follow that those things we are confident *are* unethical must be prohibited by legislation or controlled by regulation. These are separate steps that require separate arguments.

MORAL NOSE

The idea that moral sentiments, or indeed, gut reactions must play a crucial role in the determination of what is morally permissible is tenacious. This idea, originating with David Hume (who memorably remarked that morality is "more properly felt than judg'd of "), has been influential in the work of a number of contemporary moral philosophers.[8] In particular, Mary Warnock has made it a central part of her own approach to these issues. Briefly the idea is:

If morality is to exist at all, either privately or publicly, there must be some things which, regardless of consequences should not be done, some barriers which should not be passed.

What marks out these barriers is often a sense of outrage, if something is done; a feeling that to permit some practice would be indecent or part of the collapse of civilization.[9]

A recent, highly sophisticated and thoroughly mischievous example in the context of cloning comes from Leon R. Kass. In a long discussion entitled "The Wisdom of Repugnance" [see chapter 12 in this volume] Kass tries hard and thoughtfully to make plausible the thesis that thoughtlessness is a virtue:

We are repelled by the prospect of cloning human beings not because of the strangeness or novelty of the undertaking, but because we intuit and feel, immediately and without argument, the violation of things that we rightfully hold dear.[10]

The difficulty is, of course, to know when one's sense of outrage is evidence of something morally disturbing and when it is simply an expression of bare prejudice or something even more shameful. The English novelist George Orwell once referred to this reliance on some innate sense of right and wrong as "moral nose," as if one could simply sniff a situation and detect wickedness.[11] The problem, as I have indicated, is that nasal reasoning is notoriously unreliable, and olfactory moral philosophy, its theoretical "big brother," has done little to refine it or give it a respectable foundation. We should remember that in the recent past, among the many discreditable uses of so-called "moral feelings," people have been disgusted by the sight of Jews, black people, and indeed women being treated as equals and mixing on terms of equality with others. In the absence of convincing arguments, we should be suspicious of accepting the conclusions of those who use nasal reasoning as the basis of their moral convictions.

In Kass's suggestion (he disarmingly admits revulsion "is not an argument") the giveaway is in his use of the term "rightfully." How can we know that revulsion, however sincerely or vividly felt, is occasioned by the violation of things we rightfully hold dear unless we have a theory, or at least an argument, about which of the things we happen to hold dear we *rightfully* hold dear? The term "rightfully" implies a judgment that confirms the respectability of the feelings. If it is simply one feeling confirming another, then we really are in the situation Wittgenstein lampooned as buying a second copy of the same newspaper to confirm the truth of what we read in the first.

We should perhaps also note for the record that cloning was not anticipated

CLONING

by the Deity in any of his (or her) manifestations on earth; nor in any of the extant holy books of the various religions. Ecclesiastical pronouncements on the issue cannot therefore be evidence of God's will on cloning, and must be examined on the merits of the evidence and argument that inform them, like the judgments or opinions of any other individuals.

NOTES

1. For a more comprehensive account of the ethics of human cloning generally see J. Harris, "Good-bye Dolly? The Ethics of Human Cloning," *Journal of Medical Ethics* 23, no. 6 (1997).

2. A. Kahn, "Clone Mammals . . . Clone Man," *Nature* 386 (1997): 119.

3. J. Harris, "Is Cloning an Attack on Human Dignity?" *Nature* 387 (1997): 754.

4. GAEIB, *Opinion of the Group of Advisers on the Ethical Implications of Biotechnology to the European Commission,* no. 9 (May 28, 1997).

5. See note 2, Kahn 1997.

6. D. Shapiro, Letter, *Nature* 388 (1997): 511.

7. R. Winston, *British Medical Journal* 314 (1997): 913–14.

8. David Hume in his *A Treatise of Human Nature* (1738). Contemporary philosophers who have flirted with a similar approach include Stuart Hampshire; see, for example, S. Hampshire, *Morality & Pessimism: The Leslie Stephen Lecture* (Cambridge: Cambridge University Press, 1972); and Bernard Williams in "Against Utilitarianism," in B. Williams and J. C. C. Smart, *Utilitarianism For and Against* (Cambridge: Cambridge University Press, 1973). I first discussed the pitfalls of olfactory moral philosophy in my *Violence and Responsibility* (London: Routledge & Kegan Paul, 1980).

9. M. Warnock, "Do Human Cells Have Rights?" *Bioethics* 1, no. 1 (1987): 8.

10. L. R. Kass, "The Wisdom of Repugnance," *New Republic* (June 2, 1997):17–26. The obvious erudition of his writing leads to expectations that he might have found feelings prompted by more promising parts of his anatomy with which to entertain us.

11. G. Orwell, Letter to Humphrey House (April 11, 1940), in *The Collected Essays, Journalism and Letters of George Orwell,* vol. 1 (Harmondsworth: Penguin, 1970), p. 583. See my more detailed discussion of the problems with this type of reasoning in J. Harris, *Wonderwoman & Superman: The Ethics of Human Biotechnology* (Oxford: Oxford University Press, 1992).

HUMAN CLONING: SOCIETAL QUESTIONS

IX

W e are now starting to move toward some of the social issues that are raised by human cloning. Philip Kitcher's article is something of a bridge between the earlier discussions and these later ones. He is concerned about human autonomy and unlike John Harris (in the last section) responds more positively to Kant's philosophy. Kitcher denounces attempts to produce humans to order, arguing that all parents who used any kind of genetic engineering in this direction would be "demonstrating a crass failure to recognize their children as independent beings." He gives the example of John Stuart Mill force-fed Greek and Latin by his father, and of the ill effects of such a blatant attempt to produce a person to order.

However, although Kitcher does not care for attempts to design human nature, he is far from thinking that all attempts at human cloning are wrong. Like Walter Glannon (also in the last section), he wrestles with the case of children produced to order to help sick siblings, and certainly shows sympathy with the plight of parents in such a situation. There are other situations where cloning would be even more obviously justifiable. Kitcher instances such cases as people trying to recreate children who are dead or dying, and other cases where children would not be possible by normal methods of reproduction. And then there are social and personal factors to be taken into account. Kitcher finds the most justifiably permissible case of all for cloning to be that coming from the desires of lesbian lovers: they would like to have children and to this end decide to clone one partner and have the other partner bear the child.

If nothing else, in arriving at such a conclusion as this, Kitcher's discussion does show how an issue like human cloning cannot simply be decided in isolation. It is one very much dependent on societal norms and standards. One doubts very much that Kitcher's conclusion would be welcomed by Leon Kass (Section IV), or by anyone else who took an old-fashioned attitude to societal practices (such as the religious thinkers and theologians to be introduced and discussed in Sections VIII and IX). It is only with the changing attitudes to such issues as the traditional family and the moral status of nonheterosexual feelings and practices that such a suggestion as that of Kitcher is even remotely plausible.

Søren Holm is less enthused than Kitcher by human cloning. He offers what he calls a "life in the shadow argument." Anyone who is born a clone of another will spend his or her whole life in the shadow of the person who went before. All of his or her life, he or she will be compared to another person, and this will obviously bring on great psychological stress. One might of course argue that in fact the cloned person will be different—perhaps very different—from the person who was cloned. We know already (as pointed out by Stephen Jay Gould) that there is much more to being a human than genes. But, argues Holm, this is not the general perception of society. Nor is it a perception likely to change soon. "If ever the public relinquishes all belief in genetic essentialism the 'life in the shadow argument' would fail, but such a development seems highly unlikely."

Justine Burley and John Harris (again!) take up the issue of child welfare including the argument just given by Søren Holm. The two writers appreciate that a child born through cloning may well suffer societal prejudice because of his or her origins. There is the general problem that society as such may be turned against them simply because they are clones. If one disapproves very strongly of the method by which a person was produced, can one possibly feel entirely open and neutral to that person him or herself? There is indeed the problem of life in the shadow. A cloned person might go through life always being compared to the person from whom one was cloned. And there is the problem of awareness of one's own origins. Might one not feel stigmatized by one's origins, just as the child of a Nazi might feel stigmatized? Nevertheless, although Burley and Harris accept that these are real worries, they feel that the right of a person to have a cloned child outweighs the negative factors. None of the counterarguments is "sufficient to warrant state interference with the procreative choices of people who wish to clone their genes."

15 · Whose Self Is It, Anyway?

Philip Kitcher

In April 1988 Abe and Mary Ayala of Walnut, California, began living through every parent's nightmare: Anissa, their sixteen-year-old daughter, was diagnosed with leukemia. Without a bone-marrow transplant, Anissa would probably die within five years. But who could donate bone marrow that Anissa's immune system would not reject? Tests confirmed the worst: neither Abe, Mary, nor their other child had compatible marrow.

The family embarked on a desperate plan. Abe, who had had a vasectomy years before, had it surgically reversed. Within months, at the age of forty-three, Mary became pregnant. The genetic odds were still three-to-one against a match between Anissa's bone marrow and that of the unborn child. The media got hold of the story, and the unbearable wait became a public agony.

Against all the odds a healthy daughter was born with compatible bone marrow. Fourteen months later, in June 1991, physicians extracted a few ounces of the child's marrow: the elixir that would save her older sister's life.

This article is reprinted by permission of *The Sciences* and is from the September/October 1997 issue, pages 58–62. The Sciences, 2 East 63rd Street, New York, NY 10021.

CLONING

The story has a happy ending, but many people have found it at least slightly disturbing. Is it right for a couple to conceive one child to save another? Can someone brought into the world for such a well-defined purpose ever feel that she is loved for who she is? Thirty-seven percent of the people questioned in a contemporaneous *Time* magazine poll said they thought what the Ayalas had done was wrong; 47 percent believed it was justifiable.

Six years have passed and now a different, yet related, event a continent away has shaken the public's moral compass. Lamb number 6LL3, better known as Dolly, took the world by surprise last February when she was introduced as the first creature ever cloned from an adult mammal. Recognizing that what is possible with sheep today will probably be feasible with human beings tomorrow, commentators speculated about the legitimacy of cloning a Pavarotti or an Einstein, about the chances that a demerited dictator might produce an army of supersoldiers, about the future of basketball in a world where a team of Larry Birds could play against a team of Michael Jordans. Polls showed that Mother Teresa was the most popular choice for person-to-be-cloned, but the film star Michelle Pfeiffer was not far behind, and Bill and Hillary Clinton, though tainted by controversy over alleged abuses of presidential power, also garnered some support.

Beyond all the fanciful talk, Dolly's debut introduces real and pressing moral issues. Cloning will not enable anyone to duplicate people like so many cookie-cutter gingerbread men, but it will pave the way for creating children who can fulfill their parents' preordained intentions. Families in the Ayalas' circumstances, for instance, would have a new option: Clone their dying child to give birth to another whose identical genetic makeup would guarantee them a compatible organ or a tissue match. Should they be allowed to exercise that option? The ethical implications of cloning balance on a fine line.

Society can probably blame Mary Wollstonecraft Shelley and her fervent imagination for much of the brouhaha over cloning. The Frankenstein story colors popular reception of the recent news, fomenting a potent brew of associations: many people assume that human lives can be made to order, that there is something vaguely illicit about the process, and, of course, that it is all going to turn out disastrously. Reality is much more complicated—and more sobering—so one should preface debates about the morality of human cloning with a clear understanding of the scientific facts.

As most newspaper readers know by now, the recent breakthroughs in cloning did not come from one of the major centers of the genetic revolution, but from the far less glamorous world of animal husbandry and agricultural research. A team of investigators at the Roslin Institute, near Edinburgh, Scotland, led by Ian Wilmut, conjectured that past efforts to clone mammals had failed because the cell that supplied the nucleus and the egg that received it were at different stages of the cell cycle. Applying well-known techniques from cell biology, Wilmut "starved" the cells so that both were in an inactive phase at the time of transfer. Inserting nuclei from adult sheep cells in that quiescent phase gave rise to a number of embryos, which were then implanted into ewes. In spite of a high rate of miscarriage, one of the pregnancies continued to term. After beginning with 277 transferred adult nuclei, Wilmut and his coworkers obtained one healthy lamb: the celebrated Dolly.

Wilmut's achievement raises three important questions about the prospect of human cloning: Will it be possible to undertake the same operations on human cells? Will cloners be able to reduce the high rate of failure? And just what is the relation between a clone obtained through nuclear transplantation and the animals, born in the usual way, from which the clone is derived?

Answers to the first two questions are necessarily tentative; predicting even the immediate trajectory of biological research is always vulnerable to contingencies. In the late 1960s, for example, after the developmental biologist J. B. Gurdon, now of the University of Cambridge, produced an adult frog through cloning, it seemed that cloning all kinds of animals was just around the corner; a few years later, the idea of cloning adult mammals had returned to the realm of science fiction [see "The Birth of Cloning," by J. B. Gurdon, in this volume]. But leaving aside any definite time frame, one can reasonably expect that Wilmut's technique will eventually work on human cells and that failure rates will be reduced.

What about the third question, however, the relation between "parent" animal and clone? There one can be more confident. Dolly clearly has the same nuclear genetic material as the ewe that supplied the inserted nucleus. A second ewe supplied the egg into which that nucleus was inserted; hence Dolly's mitochondrial DNA came from another source. Indeed, though the exact roles played by mitochondrial DNA and other contents of the cytoplasm in vertebrate development are still unclear, one can say this much: Dolly's early development was shaped by the interaction between the DNA in the nucleus and the contents of the egg cytoplasm—the contributions of two adult females. A third sheep, the ewe into which the embryonic Dolly was implanted, provided Dolly with a uterine environment. Dolly thus has three mothers—nuclear mother, egg mother, and womb mother—and no father (unless, of course, one accords that honor to Wilmut for his guiding role).

Now imagine Holly, a human counterpart of Dolly. You might think Holly would be similar to her nuclear mother, perhaps nearly identical, particularly if the mother of the nuclear mother were also the womb mother, and if either that woman or the nuclear mother were the egg mother. Such a hypothetical circumstance would ensure that Holly and her donor shared a similar gestation experi-

CLONING

ence, as well as both nuclear and mitochondrial DNA. (Whether they would share other cytoplasmic constituents is anyone's guess, because the extent of the differences among eggs from a single donor is still unknown.)

But even if all Holly's genetic material and her intrauterine experience matched those of a single donor, Holly would not be an exact replica of that human being. Personal identity, as philosophers since John Locke have recognized, depends as much on life experiences as on genetics. Memories, attitudes, prejudices, and emotional attachments all contribute to the making of a person. Cloning creates babies, not fully formed adults, and babies mature through a series of unique events. You could not hope to ensure the survival of your individual consciousness by arranging for one of your cells to be cloned. Megalomaniacs with intimations of immortality need not apply.

Other environmental factors would also lead to differences between Holly and her donor. For one thing, the two would likely belong to different generations, and the gap in their ages would correspond to changes in educational trends, the adolescent subculture, and other aspects of society that affect children's development. Perhaps even more important, Holly and her donor would be raised in different families, with different friends, close relatives, teachers, neighbors, and mentors. Even if the same couple acted as parents to both, the time gap would change the familial circumstances.

Identical twins reared together are obviously similar in many respects, but even they are by no means interchangeable; for instance, 50 percent of male identical twins who are gay have a twin who is not [see "Whenever the Twain Meet," by Thomas J. Bouchard Jr., *The Sciences* (September/October 1997): 52]. Small differences in shared environments clearly play a large role. How much more dissimilarity, then, can be anticipated, given the much more dramatic variations that would exist between clones and their donors?

There will never be another you. If you hoped to fashion a son or daughter exactly in your image, you would be doomed to disappointment. Nevertheless, you might hope to take advantage of cloning technology to have a child of a certain kind—after all, the most obvious near-term applications for cloning lie in agriculture, where the technique could be used to perpetuate certain useful features of domestic animals, such as their capacity for producing milk, through succeeding generations. Some human characteristics are directly linked to specific genes and are therefore more amenable to manipulation—eye color, for instance. But in cloning, as in a good mystery novel, nothing is quite as simple as it seems.

Imagine a couple who are determined to do what they can to create a Hollywood star. Fascinated by the color of Elizabeth Taylor's eyes, they obtain a tissue sample from the actress and clone a young Liz. Will they succeed in creating a girl who possesses exact copies of the actress's celebrated eyes? Probably not. Small variations that occur at the cellular level during growth could modify the shape of the girl's eye sockets so that the eye color would no longer have its bewitching effect. Would the Liz clone still capture the hearts of millions? Perhaps the eyes would no longer have it.

Of course, Taylor's beauty and star appeal rest on much more than eye color. But the chances are that other physical attributes—height, figure, complexion, facial features—would also be somewhat different in a clone. Elizabeth II might overeat, for instance, or play strenuous sports, so that as a young adult her physique would be fatter or leaner than Elizabeth I's. Then there are the less tangible attributes that contribute to star quality: character and personal style. Consider what goes into something as apparently simple as a movie star's smile. Capturing as it does the interplay between physical features and personality, a smile is a trademark that draws on a host of factors, from jaw shape to sense of humor. How can anything so subtle ever be duplicated?

Fantasies about cloning Einstein, Mother Teresa, or Yo-Yo Ma are equally doomed. The traits people value most come about through a complex interaction between genotypes and environments. By fixing the genotype one can only increase the chances—never provide a guarantee—of achieving one's desired results. The chances of artificially fashioning a person of true distinction in any area of complex human activity, whether it be science, philanthropy, or artistic expression, are infinitesimal.

Although cloning cannot produce exact replicas or guarantee outstanding performance, it might be exploited to create a child who tends toward certain traits or talents. For example, had my wife and I wanted a son who would dominate the high school basketball court, we would have been ill-advised to reproduce in the old-fashioned way. At a combined height of just over eleven feet, we would have dramatically increased our chances by having a nucleus transferred from some strapping NBA star. And it is here, in the realm of the possible, that cloning scenarios devolve into moral squalor. By dabbling in genetic engineering, parents would be demonstrating a crass failure to recognize their children as independent beings with the freedom to form their own sense of who they are and what their lives mean.

Parents have already tried to shape and control their children, of course, even without the benefit of biological tools. The nineteenth-century English intellectual James Mill had a plan for his son's life, leading him to begin young John Stuart's instruction in Greek at age three and his Latin at age eight. John Stuart Mill's *Autobiography* is a quietly moving testament to the cramping effect of the life his eminent father had designed for him. In early adulthood, Mill *fils* suffered a nervous breakdown, from which he recovered, going on to a career of great intellectual distinction. But though John Stuart partly fulfilled his father's aspirations for him, one of the most striking features of his philosophical work is his passionate defense of human freedom. In *On Liberty* he writes: "Mankind are greater gainers by suffering each other to live as seems good to themselves, than by compelling each to live as seems good to the rest."

If the cloning of human beings is undertaken in the hope of generating a particular kind of person, then cloning is morally repugnant. The repugnance arises not because cloning involves biological tinkering but because it interferes with human autonomy. To discover whether circumstances might exist in which cloning would be morally acceptable, one must ask whether the objectionable motive can be removed. Three scenarios come immediately to mind.

First is the case of the dying child: Imagine a couple in a predicament similar to that of the Ayalas, which I described at the beginning of this essay. The couple's only son is dying and needs a kidney transplant within ten years. Unfortunately, neither parent can donate a compatible organ, and it may not be possible to procure an appropriate one from the existing donor pool. If a brother were produced by cloning, one of his kidneys could be transplanted to save the life of the elder son.

Second, the case of the grieving widow: A woman's beloved husband has been killed in an automobile accident. As a result of the same crash, the couple's only daughter lies in a coma with irreversible brain damage. The widow, who can no longer bear children, wants to have the nuclear DNA from one of her daughter's cells inserted into an egg supplied by another woman, so that a clone of her child can be produced through surrogate motherhood.

Third, the case of the loving lesbians: A lesbian couple wishes to have a child. Because they would like the child to be biologically connected to each of them, they request that a cell nucleus from one of them be inserted into an egg from the other, and that the embryo be implanted in the uterus of the woman who donated the egg.

No blatant attempt is made in any of these scenarios to direct the child's life; indeed, in some cases like these cloning may turn out to be morally justified. Yet lingering concerns remain. In the first scenario, and to a lesser extent in the second, the disinterested bystander suspects that children are being subordinated to the special purposes or projects of adults. Turning from John Stuart Mill to another great figure in contemporary moral theory, Immanuel Kant, one can ask whether any of the scenarios can be reconciled with Kant's injunction to "treat humanity, whether in your own person or in the person of another, always at the same time as an end and never simply as a means."

Perhaps the parents in the case of the dying child have no desire to expand their family; for them the younger brother would be simply a means of saving the really important life. And even if the parental attitudes were less callous, concerns would remain. In real case histories in which parents have borne a child to save an older sibling, their motives have been much more complex; the Ayala family seems a happy one, and the younger sister is thriving. Ironically, though, in such circumstances the parents' love for the younger child may be manifested most clearly if the project goes awry and the older child dies. Otherwise, the clone—and perhaps the parents as well—will probably always wonder whether he is loved primarily for his usefulness.

Similarly, the grieving widow might be motivated solely by nostalgia for the happy past, so that the child produced by cloning would be valuable only because she was genetically close to the dead. If so, another person is being treated as a means to understandable, but morbid, ends.

The case of the loving lesbians is the purest of the three. The desire to have a child who is biologically related to both of them is one that our society recognizes, at least for heterosexual couples, as completely natural and justifiable. There is no question in this scenario of imposing a particular plan on the nascent life—simply the wish to have a child who is the expression of the couple's mutual love. That is the context in which human cloning would be most defensible.

In recent decades, medicine has enabled many couples to overcome reproductive problems and bear their own biological children. Techniques of assisted reproduction have become mainstream because of a general belief that infertile couples have been deprived of something valuable, and that manipulating human cells is a legitimate response to their frustrations.

But do we, the members of a moral community, know what makes biological connections between parents and offspring valuable? Can we as a society assess the genuine benefits to the general welfare brought about by techniques of assisted reproduction, and do we want to invest in extending those techniques even further? Artificial insemination or in vitro fertilization could help the grieving widow and the lesbian couple in my scenarios; in both cases cloning would create a closer biological connection—but one should ask what makes that extra degree of relatedness worth striving for. As for the parents of the dying child, one can simply hope that the continuing growth of genetic knowledge will provide improved methods of transplantation. By the time human cloning is a real possibility, advances in immunology may enable patients to tolerate tissue from a broader range of sources.

Should human cloning be banned? For the moment, while biology and medicine remain ignorant of the potential risks—the miscarriages and malformed embryos that could result—a moratorium is surely justified. But what if future research on nonhuman mammals proves reassuring? Then, as I have suggested, cloning would be permissible in a small range of cases. Those cases must satisfy two conditions: First, there must be no effort to create a child with specific attributes. Second, there must be no other way to provide an appropriate biological connection between parent and child. As people reflect on the second condition, perhaps some will be moved to consider just how far medicine should go to help people have children "of their own." Many families have found great satisfaction in rearing adopted children. Although infertile couples sometimes suffer great distress, further investment in technologies such as cloning may not be the best way to bring them relief.

The public fascination with cloning reached all the way to the White House almost immediately after Wilmut's epochal announcement. President Clinton was quick first to refer the issue to his National Bioethics Advisory Commission and then to ban federal funding for research into human cloning. The response was panicky, reflexive, and disappointing. In the words of the editors of *Nature*: "At a time when the science policy world is replete with technology foresight exercises, for a U.S. president and other politicians only now to be requesting guidance about [the implications of cloning] is shaming."

But though society and its leadership are woefully unprepared to handle cloning with policies based on forethought, many people race ahead irresponsibly with fantasies and fears. Human cloning becomes a titillating topic of discussion, while policy makers ignore the pressing ethical issues of the moment. In a fit of moral myopia, the U.S. government moves to reject human cloning because of potential future ills, while it institutes policies that permit existing children to live without proper health care and that endanger children's access to food and shelter.

The respect for the autonomy of lives and the duty to do what one can to let children flourish in their own ways should extend beyond hypothetical discussions about cloning. However strongly one may feel about the plights of loving lesbians, grieving widows, or even couples with dying children, deciding how cloning might legitimately be applied to their troubles is not the most urgent moral or political question, or the best use of financial resources. I would hope that the public debate about new developments in biotechnology would ultimately spur our society to be more vigilant about applying the moral principles that we espouse but so often disregard.

Making demands for social investment seems quixotic, particularly when funds for the poor in the United States are being slashed and when other affluent countries are having second thoughts about the responsibilities of societies toward their citizens. The patronizing adjectives, such as "idealistic" and "utopian," that conservatives bestow on liberal programs do nothing to undermine the legitimacy of the demands. What is truly shameful is not that the response to the possibilities of cloning came so late, nor that the response has been so confused, but that the affluent nations have been so reluctant to think through the implications of time-honored moral principles and to design a coherent use of the new genetic science, technology, and information for human well-being.

A Life in the Shadow
One Reason Why We Should Not Clone Humans
Søren Holm

Introduction

O ne of the arguments that is often put forward in the discussion of human cloning is that it is in itself wrong to create a copy of a human being.

This argument is usually dismissed by pointing out that (a) we do not find anything wrong in the existence of monozygotic twins even though they are genetically identical, and (b) the clone would not be an exact copy of the original even in those cases where it is an exact genetic copy, since it would have experienced a different environment that would have modified its biological and psychological development.

In my view both these counterarguments are valid, but nevertheless I think that there is some core of truth in the assertion that it is wrong deliberately to try to create a copy of an already existing human being. It is this idea that I will briefly try to explicate here.

Reprinted from *Cambridge Quarterly of Healthcare Ethics* 7 (Spring 1998): 160–62. Copyright © 1998 Cambridge University Press. Reprinted with the permission of Cambridge University Press.

CLONING

The Life in the Shadow Argument

When we see a pair of monozygotic twins who are perfectly identically dressed some of us experience a slight sense of unease, especially in the cases where the twins are young children. This unease is exacerbated when people establish competitions where the winners are the most identical pair of twins. The reason for this uneasiness is, I believe, that the identical clothes could signal a reluctance on the part of the parents to let each twin develop his or her individual and separate personality or a reluctance to let each twin lead his or her own life. In the extreme case each twin is constantly compared with the other and any difference is counteracted.

In the case of cloning based on somatic cells we have what is effectively a set of monozygotic twins with a potentially very large age difference. The original may have lived all his or her life and may even have died before the clone is brought into existence. Therefore, there will not be any direct day-by-day comparison and identical clothing, but then a situation that is even worse for the clone is likely to develop. I shall call this situation "a life in the shadow" and I shall develop an argument against human cloning that may be labeled the "life in the shadow argument."

Let us try to imagine what will happen when a clone is born and its social parents have to begin rearing it. Usually when a child is born we ask hypothetical questions like "How will it develop?" or "What kind of person will it become?" and we often answer them with reference to various psychological traits we think we can identify in the biological mother or father or in their families, for instance "I hope that he won't get the kind of temper you had when you were a child!"

In the case of the clone we are, however, likely to give much more specific answers to such questions. Answers that will then go on to affect the way the child is reared. There is no doubt that the common public understanding of the relationship between genetics and psychology contains substantial strands of genetic essentialism, i.e., the idea that the genes determine psychology and personality.[1] This public idea is reinforced every time the media report the finding of new genes for depression, schizophrenia, etc. Therefore, it is likely that the parents of the clone will already have formed in their minds a quite definite picture of how the clone will develop, a picture that is based on the actual development of the original. This picture will control the way they rear the child. They will try to prevent some developments, and try to promote others. Just imagine how a clone of Adolf Hitler or Pol Pot would be reared, or how a clone of Albert Einstein, Ludwig van Beethoven, or Michael Jordan would be brought up. The clone would in a very literal way live his or her life in the shadow of the life of the original. At every point in the clone's life there would be someone who had already lived that life, with whom the clone could be compared and against whom the clone's accomplishments could be measured.

That there would in fact be a strong tendency to make the inference from genotype to phenotype and to let the conclusion of such an inference affect rearing can perhaps be seen more clearly if we imagine the following hypothetical situation:

> In the future new genetic research reveals that there are only a limited number of possible human genotypes, and that genotypes are therefore recycled every 300 years (i.e., somebody who died 300 years ago had exactly the same genotype as me). It is further discovered that there is some complicated, but not practically impossible, method whereby it is possible to discover the identity of the persons who 300, 600, 900, etc. years ago instantiated the genotype that a specific fetus now has.

I am absolutely certain that people would split into two sharply disagreeing camps if this became a possibility. One group, perhaps the majority, would try to identify the previous instantiations of their child's genotype. Another group would emphatically not seek this information because they would not want to know and would not want their children to grow up in the shadow of a number of previously led lives with the same genotype. The option to remain in ignorance is, however, not open to social parents of contemporary clones.

If the majority would seek the information in this scenario, firms offering the method of identification would have a very brisk business, and it could perhaps even become usual to expect of prospective parents that they make use of this new possibility. Why would this happen? The only reasonable explanation, apart from initial curiosity, is that people would believe that by identifying the previous instantiation of the genotype they would thereby gain valuable knowledge about their child. But knowledge is in general only valuable if it can be converted into new options for action, and the most likely form of action would be that information about the previous instantiations would be used in deciding how to rear the present child. This again points to the importance of the public perception of genetic essentialism, since the environment must have changed considerably in the three-hundred-year span between each instantiation of the genotype.

WHAT IS WRONG ABOUT A LIFE IN THE SHADOW?

What is wrong with living your life as a clone in the shadow of the life of the original? It diminishes the clone's possibility of living a life that is in a full sense of that word his or her life. The clone is forced to be involved in an attempt to perform a complicated partial reenactment of the life of somebody else (the original). In our usual arguments for the importance of respect for autonomy or for the value of self-determination we often affirm that it is the final moral basis for these principles that they enable persons to live their lives the way they themselves want to live these lives. If we deny part of this opportunity to clones and force them to live their lives in the shadow of someone else we are violating some of our most fundamental moral principles and intuitions. Therefore, as long as genetic essentialism is a common cultural belief there are good reasons not to allow human cloning.

Final Qualifications

It is important to note that the "life in the shadow argument" does not rely on the false premise that we can make an inference from genotype to (psychological or personality) phenotype, but only on the true premise that there is a strong public tendency to make such an inference. This means that the conclusions of the argument only follow as long as this empirical premise remains true. If ever the public relinquishes all belief in genetic essentialism the "life in the shadow argument" would fail, but such a development seems highly unlikely.

In conclusion I should perhaps also mention that I am fully aware of two possible counterarguments to the argument presented above. The first points out that even if a life in the shadow of the original is perhaps problematic and not very good, it is the only life the clone can have, and that it is therefore in the clone's interest to have this life as long as it is not worse than having no life at all. The "life in the shadow argument" therefore does not show that cloning should be prohibited. I am unconvinced by this counterargument, just as I am by all arguments involving comparisons between existence and nonexistence, but it is outside the scope of the present short paper to show decisively that the counterargument is wrong.

The second counterargument states that the conclusions of the "life in the shadow argument" can be avoided if all clones are anonymously put up for adoption, so that no knowledge about the original is available to the social parents of the clone. I am happy to accept this counterargument, but I think that a system where I was not allowed to rear the clone of myself would practically annihilate any interest in human cloning. The attraction in cloning for many is exactly in the belief that I can recreate myself. The cases where human cloning solves real medical or reproductive problems are on the fringe of the area of cloning.

NOTE

1. D. Nelkin and M. S. Lindee, *The DNA Mystique: The Gene as a Cultural Icon* (New York: W. H. Freeman and Company, 1995).

17 Human Cloning and Child Welfare
Justine Burley and John Harris

INTRODUCTION

D ebate over the moral permissibility of human cloning was much enlivened by the news that the first mammal, Dolly the sheep, had been successfully created from the transfer of an already differentiated adult cell nucleus into an enucleated egg,[1] a technique popularly referred to as cloning. Now that doubts about Dolly's genetic origins have, for the most part, been dispelled,[2] and that three other species, including the mouse,[3] have been cloned, the prospect that a human may also be cloned appears ever more likely.[4]

There is a broad, albeit loose, consensus among members of the lay public, various legislative bodies, and the scientific community that human *reproductive*[5] cloning should be banned because there is something deeply immoral about it *in principle* (i.e., something above and beyond the fact that it would be wholly unac-

Reprinted with permission of BMJ Publishing Group from *Journal of Medical Ethics* 25 (1999): 108–13.

ceptable to attempt it in humans until it appears reasonably safe). Precisely what this is, however, continues to elude even the most committed of critics.[6] These opponents are not short of reasons for their anti-human cloning stance, indeed, such reasons, more often than not couched in mysterious appeals to human rights and human dignity, flow freely.[7] But none of the objections to the practice of human cloning have so far proved sound or convincing. In this paper we address one of the few intelligible (as opposed to persuasive) objections to human cloning that have been advanced: the objection which appeals to the welfare of the child. The form of this objection varies according to the sort of harm it is anticipated the clone will suffer. The three formulations of it that we will consider are:

1. Clones will be harmed by the fearful or prejudicial attitudes people may have about or towards them (H1);
2. Clones will be harmed by the demands and expectations of parents or genotype donors (H2);
3. Clones will be harmed by their own awareness of their origins, for example the knowledge that the genetic donor is a stranger (H3).

Below we aim to show why these three versions of the child welfare objection do *not* supply compelling reasons to ban human reproductive cloning. The claim that we will develop and defend in the course of our discussion is that even if it is the case that a cloned child will suffer harms of the type H1–H3, it is none the less permissible to conceive by cloning so long as these cloning-induced welfare deficits are not such as to blight the existence of the resultant child, whoever this may be.

Our article is divided into four main parts. We begin by outlining what Derek Parfit has called the "nonidentity problem."[8] As we will demonstrate, this problem, when explored and understood properly, shows that those who object to human cloning on the grounds that it would have compromising effects on a *particular* child's welfare are making an error in reasoning. We then go on to outline Derek Parfit's principled solution to the nonidentity problem and, in the ensuing three sections, we will argue against each of the three objections to human cloning from child welfare identified above, in turn, by reference to this solution. Once it is seen that Parfit's principle in unqualified form is what informs these objections it becomes clear that people who object to human cloning for reasons of child welfare are impaled on the horns of a dilemma: either they must concede that their position entails a whole host of morally unpalatable restrictions on both artificial and natural procreation or they must accept that their arguments are insufficient to support the view that human cloning is immoral in any strong sense and so should be prohibited.

CLONING

I. Child Welfare and the Nonidentity Problem

One of the chief philosophical problems raised by human cloning is the question of how we should respond to the interests of people not yet in existence. Objections to human cloning from child welfare are objections relating to harms future clones might come to suffer and they may all be captured by the claim: a child who is cloned would, *for that reason or for reasons related but not intrinsic to it,* suffer a deficit in well-being relative to someone conceived through natural means.

Typical in discussions of this claim is the notion that the harms and benefits which concern us, occur to the same child. However, as exploration of the nonidentity problem will now make clear, claims of this kind cannot explain what it is that might be thought problematic about a decision which results in a clone being harmed in any way at all.

To give shape to the nonidentity problem we will now consider the following two cases. The first is Parfit's and involves a fourteen-year-old prospective mother:

> This girl chooses to have a child. Because she is so young, she gives her child a bad start in life. Though this will have bad effects throughout the child's life, his or her life will, predictably, be worth living. If this girl had waited for several years, she would have had a different child, to whom she would have given a better start in life.[9]

Our analogue to this case is:

> A woman chooses to have a child through cloning. Because she chooses to conceive in this way, she gives the child a bad start in life. Though this will have bad effects throughout the child's life, his or her life will, predictably, be worth living. If this woman had chosen to procreate by alternative means, she would have had a different child, to whom she would have given a better start in life.

In both cases, two courses of action are open to the prospective mother. In criticizing these women's pursuit of the first option available (i.e., conception at fourteen and reproductive cloning respectively) people might claim that each mother's decisions will probably be worse for *her child.*[10] However, as Parfit notes, while people can make this claim about the decisions taken it does not *explain* what they believe is objectionable about them. It fails to explain this because neither decision can be worse for the particular children born; *the alternative for both of them was never to have existed at all.* If the fourteen-year-old waits to conceive, a completely different child will be born. Likewise, if the woman chooses not to clone and instead conceives by natural procreative means the child born will be a completely different one. Thus claims about the badness of pursuing the first option in both of the above cases cannot be claims about why *these* children have been harmed. It is better for *these* children that they live than not live at all.

Parfit's solution to the nonidentity problem is to posit claim "Q," which says that: "If in either of two possible outcomes the same number of people would ever live, it would be worse if those who live are worse off, or have a lower quality of life, than those who would have lived."[11] This claim, unlike the claim about the welfare of a particular child, can explain the goodness and badness of the procreative decisions that might be taken by the two women in the above cases because it avoids the problem of nonidentity.

With respect to the two cases we have been examining Q implies that the fourteen-year-old girl should wait and that the woman should not use cloning technology. This is not necessarily Parfit's last word on the matter. He does qualify Q: he argues that some things may matter more than suboptimal outcomes.[12] For example, a society might believe that the pursuit of equality is more valuable than promoting economic growth. Parfit, then, is a pluralist. He believes that Q is a helpful principle with which to evaluate moral judgments, but he does not think that this principle should necessarily be used to the exclusion of all others. It is, however, something very like Parfit's principle in its unqualified form (henceforth Q(U)) to which those who object to human cloning from child welfare are appealing, i.e., the idea that the principle factor that should weigh in decision making about who should be brought into existence is the question of who will enjoy the highest level of welfare. We will now make explicit why we believe that this approach is gravely mistaken through an examination of objections 1–3 as stated above.

II. Cloning, Societal Prejudice

The first objection to human cloning from child welfare we will address says that human cloning should be disallowed because clones will be harmed by the fearful and/or prejudicial attitudes other people have about or towards them (H1).[13] The chairman of the Human Fertilisation and Embryology Authority (HFEA), Ruth Deech, offers paradigmatic examples of this objection in a recent comment on the subject:

> Would cloned children be the butt of jibes and/or be discriminated against? Would they become a subcaste who would have to keep to each other? Would they be exploited? Would they become media objects (not an unlikely scenario

given that Louise Brown, the first test-tube baby, is still in the media some twenty years after her conception)?[14]

Deech's objection here gives primacy to the well-being of future clones. Cloning is thought undesirable because of H1 type harms that they might suffer. Her view appears to be informed by Q(U), according to which, the decision to clone should not be taken as any resulting child, other things being equal, would have a worse life than a child produced by natural means.

But, it is utterly crucial that we do not lose sight of the reason why in this case the clone's life would be the worse one, namely, that other members of society are prejudiced against him or her. Q(U) as applied by Deech entails morally repugnant conclusions. Deech's deployment of Q(U) reasoning does not show that parents who chose to clone would be acting immorally. The source of the harm is not the clone's parents, it is not they who do something wrong by cloning the child, rather it is other members of society who commit a moral wrong. Think of interracial marriage in a society hostile to mixed-race unions.

The following example involving Tom, Dick, and Harry illustrates our point here. Suppose that a woman called Jane can conceive a child with either Tom, Dick, or Harry, and that for her (not others) the only relevant difference between these men is that Tom, predictably, will be the better father; if either Dick or Harry were, predictably, the better father, she would choose one of them as her mate instead. Deech is committed to saying that Jane ought to choose Tom, as, if chosen, the resulting child will, other things being equal, have a higher level of well-being than any child parented by Dick or Harry. Let us more fully describe this scenario. Assume that the reasons Tom will be the "better" father is that the society in which all four of these individuals live is predominately populated by white people, a fair number of whom are racist and that of Tom, Dick, and Harry, only Tom is white. He is the "better" father because, in this racist society his skin color and cultural background afford him better employment and other life-enhancing opportunities and therefore he is better able to provide for any child that he and Jane (also white) conceive. Moreover, it is the case that if Jane selects Tom, they will not have a mixed-race child (as would occur were she to pair up with Dick or Harry instead), and therefore, because of the prevailing climate of racial prejudice any child born to Jane and Tom will lead a better life than any child born to Jane and Dick or Jane and Harry.

Q(U) recommends Jane's choice of Tom because the children Jane might have with Dick or Harry would lead worse lives *because of the prejudice of others*. However, to reject cloning on the grounds of this variant of the objection from child welfare is morally discreditable. It is true that we could prevent this sort of harm being done to a future child by avoiding human cloning altogether. But we should not prevent human cloning in the face of this sort of suboptimality, rather we should concentrate on combating the prejudices and attitudes that are the source of harm to the clone. Those who embrace liberty and respect autonomy will prefer this approach and reject assaults on human freedom and dignity of the kind Deech suggests would be perpetrated on clones. Plainly, it is inappropriate to countenance any diminution in reproductive autonomy when attempts to diminish prejudice and tyranny are all the more consonant with human dignity.

III. CLONING, "LIFE IN THE SHADOW"

It has been claimed by a number of critics that a clone might be harmed because of the expectations and demands of his or her parents or genotype donor (H2) and therefore reproductive cloning should be proscribed. Søren Holm, for example, argues that one reason that we have not to clone a human is that the clone will be living "a life in the shadow" of the person from whose genes he or she was cloned— the clone would not have a life that was fully his or her own.[15] Holm's argument may be stated in brief as follows: people are wedded to a belief in genetic essentialism (i.e., they misunderstand the relationship between genetics and personality), for that reason a clone's parent(s) may treat him or her such that the clone will not lead an autonomous life, he or she will always be *living in the shadow* of another (i.e., the genetic donor)—incessantly compared to the donor—therefore human cloning should not be allowed.[16] It is unclear from what Holm has said what, specifically, he believes the relationship between autonomy and well-being to be. If, like prominent liberal thinkers such as Ronald Dworkin, he thinks that autonomy is *part* of well-being then his objection to human cloning is an objection about the welfare of the child. If, on the other hand, he, like Kant, understands autonomy as an independent principle then the objection is not, per se, one about the welfare of the child. We shall respond to Holm as though he is advancing the former sort of objection.[17]

Note that the claim on which Holm's argument is premised (he calls it the "true" premise) is that the public harbor misunderstandings about genetic essentialism, i.e., they make a factual error. It is this crucial factual premise in Holm's argument which undermines its major normative force. Holm concedes that were the public to be disabused of its views about genetic essentialism the life-in-the-shadow argument would fall flat.[18] But, he insists, such a change in public understanding about genetics is unlikely.[19] Apart from the fact that we do not share Holm's dim view of what the lay public is likely to understand about genetics, the

CLONING

life-in-the-shadow argument is lacking in a different important way. It is morally problematic to limit human freedom on the basis of false beliefs of this character. Were we to apply the logic of Holm's argument to other factual errors parents might make or false beliefs they might have which would affect the wellbeing of possible children it would have pernicious implications. For example, parents might falsely believe that certain physical deformities implied intellectual impairment and this would lead them to treat children so deformed in a way which undermined their autonomy. Should such people be denied the freedom to procreate whenever it was known that they might conceive such a child? Likewise, parents might believe that female children were less intelligent than males and, in grooming them for marriage from birth, deny them an autonomous existence. Should such parents only be allowed to have male offspring? Holm's argument against human cloning appears also to commit him to restrictive procreative policies like these which undoubtedly would adversely impact, for the most part, on people who are ill-educated/ill-informed (or genetically unlucky). If human cloning is banned because future people might suffer harms caused by the mistaken beliefs of parents about genetics then it follows that so too might natural procreation whenever prospective parents do not possess adequate factual information to ensure any future child's well-being in other ways. We reject this conclusion and propose that the preferable strategy for dealing with the problem Holm highlights is one which involves educating people about genetics.

Holm rightly signals that the moral basis for arguments about respect for autonomy is a claim about the fundamental importance or value of having control over the pursuit of one's own projects, plans, and attachments. The ideal of autonomy is used by liberal theorists to defend a particular role for the state, namely, the creation and maintenance of the social, economic, and political conditions under which people may learn about different aspects and ways of life, reflect critically on them, and embrace a set of values and aims which they believe give life meaning. While it may be true that the autonomy of a clone who lived her life in the shadow of another would be adversely affected this is not sufficient to curtail a would-be cloner's reproductive freedom. Holm fails adequately to appreciate that the liberal ideal of autonomy to which he appeals requires, amongst other things, that compelling reasons (construed as reasons which squarely locate a deeply immoral outcome) must be given to limit individual freedom. Holm champions impediments to autonomous living as a sufficient reason to ban cloning but he is surely mistaken.[20] In failing to distinguish between his idea of a clone living a life in the shadow and the degree of the harm which that entails from other acts of procreation involving equally, if not more severe, autonomy-affecting consequences, Holm invites highly illiberal restrictions on procreation. Freedom is costly—affording it to individuals will, in many cases, produce suboptimal outcomes but unless these outcomes involve a moral wrong so serious that freedom must be sacrificed to prevent it, the liberal view insists that freedom prevail and that other means be found to combat any resulting harms.

Holm's objection to human cloning is more sophisticated than the one we considered in II above. It implies that parents who elect to clone, who do not understand the distinction between genotype and phenotype, are committing a moral wrong of some kind. But how serious is this moral wrong? What Q(U) reasoning in this case shows is that this principle is useful conditionally on the severity of the harm inflicted. We maintain that unless it is shown convincingly that "living in the shadow" is somehow both horrendous and more autonomy-compromising than the plethora of other widely accepted and permitted upbringings a child might be "forced" to undergo, the liberal principle of freedom in matters relating to procreation overrides the concern about autonomy-related welfare deficits that will be suffered by clones.

IV. CLONING AND AWARENESS OF ORIGINS

The final objection to human cloning from child welfare we shall explore concerns instances of psychological harms caused by a clone's own awareness of his or her genetic origins (H3). According to this objection, a clone who knew his or her genetic donor was, for example, a randomly chosen stranger, or a distant, much older relative, or even someone now deceased, would be psychologically damaged by that information.[21] Is this plausible? We doubt that knowledge of peculiar genetic origins would *necessarily* be harmful. Indeed, it might even be beneficial in certain cases. In making this claim we have in mind children who are the product of in vitro fertilization (which need not always involve the genetic material of both the parents), who report that they feel "special" (as opposed to alienated) for having been brought into being in this way. Presumably this has much to do with the extent to which they feel loved by the parents they do have, as well as societal acceptance of IVF as a procreative method.

However, let us assume that it would be the case that a clone would be traumatized to some extent by his or her genetic origins. Is this sufficient reason to disallow cloning? If H3 harms are both very great and highly probable then, yes, this is a sufficient reason; but we judge this scenario to be a remote possibility. Consider that there are many possible sources of analogous H3 type traumas a child created by natural means might suffer: the realization that your parent committed a criminal act earlier in his/her life, or is a drug addict or prostitute, or fought for an army established to advance a dictator's master plan for domination. Our intuition is that it would be far easier to cope with the knowledge that one's nurturing parents so desired a child that they were even prepared to use cloning technology to bring one into existence than to cope with the knowledge of, for example, a parent's collusion with the Nazis' systematic extermination of the Jews or Stalin's political reeducation programs in the Siberian gulags. These examples are admittedly provocative, but they are not isolated ones, and that is the point that merits stressing. If psychological distress about one's genetic origins is sufficient to ban cloning then it follows that people who fall into the aforementioned groups and others ought also to be (or to have been) prevented from procreating.

Thus undiscriminating adherence to Q(U) reasoning invites the response that this objection from child welfare, like the preceding two, logically entails other draconian restrictions on procreative freedom which the objector would hardly endorse. Even if such critics were prepared to go that far, their view should not be tolerated in any society which aims to promote freedom of the individual. Most people believe, and they do so rightly, that we should be concerned about the sorts of lives that future people will lead, but that, at the same time, this concern should not be our sole one. If H3 harms were of exclusive import, we would have grounds for saying that a huge number of people in the world today are morally blameworthy in some strong sense for having brought children into the world.

Conclusion

We argued above that the objections that have been voiced about human cloning and child welfare are misleading. While we are sympathetic to what motivates them—society both does and should have an interest in the well-being of future people—we do not believe that the formulations of the anticloning arguments from child welfare that we have addressed are persuasive. We conceded that cloned individuals might indeed suffer welfare deficits (relative to a nonclone) but argued that even the likely occurrence of them is not sufficient to warrant state interference with the procreative choices of people who wish to clone their genes (or those of others, providing consent to their use in this way has been given).

Our examination of the objections 1–3 which respectively embody reference to harms H1–H3 are informed, we have suggested, by something very like Parfit's solution to the nonidentity problem in unqualified form. Our discussion of these objections confronts those who object to human cloning for reasons of child welfare with a dilemma: either they must endorse the morally discreditable outcomes entailed by the principle guiding their view or they must admit that Q(U), as they have deployed it, does not provide sufficient reasons for branding reproductive cloning immoral either at all or in any strong sense of that term.

Where considerations of the welfare of the child are invoked in reproduction, including in the case of reproductive cloning, we need constantly to bear in mind the following questions and the distinctions they encapsulate: is it clear that the child who may result from cloning will be so adversely affected that it will be seriously wronged by the decision? Or rather is it the case that we have general anxieties about the likely advantages and disadvantages of being cloned, for example, that disincline us to look on it with much favor? Where it is rational to judge that an individual would not have a worthwhile life if he or she were to be brought into being in particular circumstances, then we have not only powerful reasons not to make such choices ourselves but also powerful moral reasons for preventing others from so doing if we can, by legislation or regulation if necessary. However, where we judge the circumstances of a future person to be less than ideal but not so bad as to deprive that individual of a worthwhile existence, then we lack the moral justification to impose our ideals on others. The difference we are looking for is the difference between considerations which would clearly blight the life of the resulting child, and considerations that would merely make existence suboptimal in some sense. We may be entitled to prevent people from acting in ways which will result in blighted lives. We are surely on less firm, and less clearly morally respectable, grounds when we attempt to impose our ideals and preferences about the specifics of how future lives should go.

Q(U), as we have shown, has troublesome practical implications for a whole range of policies concerning procreation, both natural and artificial. We argued that the reasons why a future clone's, or, for that matter, a future nonclone's life might go badly (relative to someone else), command attention. If we allow considerations like marginalization, discrimination, impediments to autonomy, etc., to outweigh all other considerations when deliberating over the moral permissibility of human cloning, we, at the same time, court numerous other unacceptably illiberal outcomes. There are, of course, many cases where it is true that the morally superior of two otherwise identical procreative acts will be the one that maximizes child welfare. The crucial issue is what follows from this. Many people believe that the child welfare card trumps all, that once they have shown that some procreative choice or technology can lead to suboptimal circumstances for the resulting children this constitutes a knock-down argument against any claimed freedom to procreate in that way or using that technology. This seems to us not only implausible but palpably morally unacceptable.

Acknowledgments

We are grateful to Matthew Clayton, G. A. Cohen, Julian Savulescu, and two anonymous referees from the *Journal of Medical Ethics* for their helpful comments on an earlier draft of this article.

References and Notes

1. I. Wilmut, A. E. Schnieke, J. McWhir, A. J. Kind, and K. H. S. Campbell, "Viable Offspring Derived from Fetal and Adult Mammalian Cells," *Nature* 385 (1987): 810–13.

2. See D. Ashworth, M. Bishop, K. Campbell, A. Colman, A. Kind, A. Schnieke, et al., "DNA Microsatellite Analysis of Dolly," *Nature* 394 (1998): 329; and E. N. Signer, Y. E. Dubrova, A. J. Jeffreys, C. Wilde, L. M. B. Finch, M. Wells, et al., "DNA Fingerprinting Dolly," *Nature* 394 (1998): 330.

3. T. Wakayama, A. C. F. Perry, M. Zuccottis, K. R. Johnson, and R. Yanagimachi, "Full-term Development of Mice from Enucleated Oocytes Injected with Cumulus Cell Nuclei," *Nature* 394 (1998): 369–73. This development in nuclear transfer technology is significant for the case of human cloning because mice possess a reproductive physiology closer to that of human beings than animals like sheep and cows.

4. This view is echoed in the opinion section of *Nature* 394 (1998).

5. Two main uses of cloning by nuclear transfer have been distinguished: therapeutic and reproductive. Therapeutic cloning is understood as any instance of cell nucleus replacement aimed at creating cell lines and/or for the treatment of disease. Reproductive cloning, by contrast, is any instance of cloning which is not motivated by the desire to avoid disease or disability. We have chosen to concentrate on reproductive uses of nuclear transfer technology because they are more controversial.

6. For discussion of why it may be difficult to pinpoint the source of people's discomfort with human cloning see L. R. Kass, "The Wisdom of Repugnance," *New Republic* (June 2, 1997): 17–26.

7. For a critique of these see J. Harris, "Good-bye Dolly? The Ethics of Human Cloning," *Journal of Medical Ethics* 23 (1997): 353–60.

8. D. Parfit, *Reasons and Persons* (Oxford: Clarendon Press, 1984), ch. 16.

9. See reference 8, p. 358.

10. This is Parfit's point made here in the plural. Reference 8, p. 359.

11. See reference 8, p. 360.

12. See Parfit's discussion of Jane's choice: reference 8, pp. 375–77.

13. We do *not* mean by the term prejudicial attitudes here *formal discrimination*, i.e., rights violations.

14. R. Deech, "Human Cloning and Public Policy," in *The Genetic Revolution and Human Rights*, ed. J. Burley (Oxford: Oxford University Press, in press), ch. 4.

15. S. Holm, "A Life in the Shadow: One Reason Why We Should Not Clone Humans," *Cambridge Quarterly of Healthcare Ethics* 7 (1998): 160–62. Other formulations of the objection may be found in: I. Wilmut, "Dolly: The Age of Biological Control," in Burley, *The Genetic Revolution and Human Rights*, ch. 1 (see reference 14); and A. J. Klotzko, "Voices from Roslin: The Creators of Dolly Discuss Science, Ethics and Social Responsibility," *Cambridge Quarterly of Healthcare Ethics* 7 (1998): especially, 137–39.

16. See reference 15: Holm, "A Life in the Shadow," p. 160.

17. Holm's own remarks suggest that this is appropriate. See reference 15: Holm, "A Life in the Shadow," p. 162.

18. Ibid.

19. Ibid.

20. Matthew Clayton has developed an ingenious argument in support of the claim that the very act of choosing the genes of a child, irrespective of the consequences, is, in a non-person-affecting sense, a violation of its autonomy. See "Procreative Autonomy and Genetics," in *A Companion to Genethics*, ed. J. Burley and J. Harris (Oxford: Blackwell, forthcoming 1999).

21. Deech, "Human Cloning and Public Policy," ch. 4.

Human Cloning: Medical Implications

What about all of the medical benefits that we have seen promised or hinted at in previous articles? A short op-ed piece by Robert Winston in the *British Medical Journal*, just after Dolly was announced, gives a sense of the excitement produced by the prospects of cloning, especially human cloning. "Not a moral threat but an exciting challenge." Which is all very well and understandable. But sometimes first reactions can be ill-thought and wrongly directed. What can one say at a more reflective pace?

Julian Savulescu argues that we have a positive moral obligation to use cloning to provide cells, tissues, and even organs for therapeutic purposes. He offers a sliding scale of examples, starting with the obviously morally acceptable case where you give a dying man a drug to stimulate the growth of healthy body cells, then taking us through more and more complex examples to the point where cloning is an essential part of the process. Savulescu argues that there is no moral divide along this scale: that indeed this all shows that there are times when we must use cloning or risk moral failure. Then building on this, he argues that cloning gives us a new perspective on conception, and that with cloning, all cells are potentially new persons. Henceforth, we can no longer believe in the special status of a fertilized egg. Many of the old arguments about the distinctive nature of the developing zygote are now otiose, and we should feel free to make use as we wish of such products of modern technology. "Producing embryos and early fetuses as a source of tissue for transplantation may be morally obligatory."

Savulescu appreciates that many will find his position unacceptable. He touches on the kind of argument that Kass finds so persuasive, namely our sense of moral revulsion at certain practices, which may include the harvesting of body parts from developing fetuses. However this, Savulescu feels, is no basis for a genuine moral argument. There are lots of things that people find disgusting when they first encounter them—artificial insemination, for instance. This does not mean that they are genuinely immoral. "The achievement in applied ethics, if there is one, in the last fifty years has been to get people to rise above their gut feelings and examine the reasons for a practice." So it is with cloning.

Robert Williamson disagrees with Savulescu, and with Justine Burly and John Harris from the last section. He brings us back to the uniqueness argument. He thinks it is obviously unethical to clone one hundred thousand people from a single cell or the cells of a single person. Hence it is unethical to clone a single person from another's cell. Why? Because there is "a personal right, ethically based, to individuality, autonomy, and identity." Moreover, specifically against Savulescu, he argues that however obtained there is a special standing or status to a developing individual human. There is a "fundamental ethical barrier" which is crossed in the development of the human—somewhere between three weeks when the primitive brain appears and thirty weeks when the fetus can survive outside the womb, and this means that we cannot treat embryos as objects to be harvested. Therefore, we must respect the "ethical value" of human "genetic uniqueness at conception."

The Promise of Cloning for Human Medicine

18

Robert Winston

THE SCIENTIFIC BACKGROUND

The production of a sheep clone, Dolly, from an adult somatic cell[1] is a stunning achievement of British science. It also holds great promise for human medicine. Sadly, the media have sensationalized the implications, ignoring the huge potential of this experiment. Accusations that scientists have been working secretively and without the chance for public debate are invalid. Successful cloning was publicized in 1975,[2] and it is over eight years since Prather et al. published details of the first piglet clone after nuclear transfer.[3]

Missing from much of the debate about Dolly is recognition that she is not an identical clone. Part of our genetic material comes from the mitochondria in the cytoplasm of the egg. In Dolly's case only the nuclear DNA was transferred. Moreover, we are a product of our nurture as much as our genetic nature.

Reprinted with permission of BMJ Publishing Group from *British Medical Journal* 314 (March 29, 1997): 913.

CLONING

Monovular twins are genetically closer than are artificially produced clones, and no one could deny that such twins have quite separate identities.

Dolly's birth provokes fascinating questions. How old is she? Her nuclear DNA gives her potentially adult status, but her mitochondria are those of a newborn. Mitochondria are important in the aging process because aging is related to acquired mutations in mitochondrial DNA, possibly caused by oxygen damage during an individual's life.[4] Experimental nuclear transfer in animals and in human cell lines could help elucidate mechanisms for many of these processes.

Equally extraordinary is the question concerning the role of the egg's cytoplasm in mammalian development. Once the quiescent nucleus had been transferred to the recipient egg cell, developmental genes expressed only in very early life were switched on. There are likely to be powerful factors in the cytoplasm of the egg that make this happen. Egg cytoplasm is perhaps the new royal jelly. Studying why and how these genes switch on would give important information about both human development and genetic disease.

Research on nuclear transfer into human eggs has immense clinical value. Here is a model for learning more about somatic cell differentiation. If, in due course, we could influence differentiation to give rise to targeted cell types we might generate many tissues of great value in transplantation. These could include skin and blood cells, and possibly neuronal tissue, for the treatment of injury, for bone marrow transplants for leukemia, and for degenerative diseases such as Parkinson's disease. One problem to be overcome is the existence of histocompatibility antigens encoded by mitochondrial DNA,[5] but there may be various ways of altering their expression. Cloning techniques might also be useful in developing transgenic animals—for example, for human xenotransplantation.

There are also environmental advantages in pursuing this technology. Mention has been made of the use of these methods to produce dairy herds and other livestock. This would be of limited value because animals with genetic diversity derived by sexual reproduction will always be preferable to those produced asexually. The risk of a line of farm animals prone to a particular disease would be ever present. However, cloning offers real prospects for preservation of endangered or rare species.

In human reproduction, cloning techniques could offer prospects to sufferers from intractable infertility. At present there is no treatment, for example, for those men who exhibit total germ cell failure. Clearly it is far-fetched to believe that we are now able to reproduce the process of meiosis, but it may be possible in future to produce a haploid cell from the male which could be used for fertilization of female gametes. Even if straight cloning techniques were used, the mother would contribute important constituents—her mitochondrial genes, intrauterine influences, and subsequent nurture.

Regulation of cloning is needed, but British law already covers this. Talk of "legal loopholes"[6] is wrong. The Human Fertilization and Embryology Act may need modification, but there is no particular urgency. A precipitate ban on human nuclear transfer would, for example, prevent the use of in vitro fertilization and preimplantation diagnosis for those couples at risk of having children who have appalling mitochondrial diseases.[7] Self-regulation and legislation already work well. Apart from any other consideration, it seems highly unlikely that doctors would transfer human clones to the uterus out of simple self-interest. Many of the animal clones that have been produced show serious developmental abnormalities,[8] and, apart from ethical considerations, doctors would not run the medicolegal risks involved. Transgenic technology has been with us for twenty years, but no clinician has been foolish enough to experiment with human germ cell therapy. The production of Dolly should not be seen as a moral threat, but rather as an exciting challenge. To answer this good science with a knee-jerk political reaction, as did President Clinton recently,[9] shows poor judgment. In a society which is still scientifically illiterate, the onus is on researchers to explain the potential good that can be gained in the laboratory.

NOTES

1. I. Wilmut, A. K. Schnieke, J. McWhir, A. J. Kind, and K. H. S. Campbell, "Viable Offspring Derived from Fetal and Adult Mammalian Cells," *Nature* 385 (1997): 810–13 [Medline].

2. J. B. Gurdon, R. A. Laskey, and O. R. Reeves, "The Developmental Capacity of Nuclei Transplanted from Keratinized Skin Cells of Adult Frogs," *Journal of Embryology and Experimental Morphology* 34 (1975): 93–112.

3. R. S. Prather, M. M. Simms, and N. L. First, "Nuclear Transplantation in Early Pig Embryos," *Biology of Reproduction* 41 (1989): 414–18.

4. T. Ozawa, "Mitochondrial DNA Mutations Associated with Aging and Degenerative Diseases," *Experimental Gerontology* 30 (1995): 269–90.

5. V. M. Dabhi and K. F. Lindahl, "MtDNA-encoded Histocompatibility Antigens," *Methods in Enzymology* 260 (1995): 466–85.

6. E. Masood, "Cloning Technique 'Reveals Legal Loophole,'" *Nature* 385 (1997): 757 [Medline].

7. R. M. Winston and A. H. Handyside, "New Challenges in Human In Vitro Fertilization," *Science* 260 (1993): 932–36.

8. K. H. S. Campbell, J. McWhir, W. A. Ritchie, and I. Wilmut, "Sheep Cloned by Nuclear Transfer from a Cultured Cell Line," *Nature* 380 (1996): 64–66.

9. J. Wise, "Sheep Cloned from Mammary Gland Cells," *British Medical Journal* 314 (1997): 623 [Full text].

Should We Clone Human Beings?
Cloning as a Source of Tissue Transplantation
Julian Savulescu

INTRODUCTION

W hen news broke in 1997 that Ian Wilmut and his colleagues had successfully cloned an adult sheep, there was an ill-informed wave of public,
professional, and bureaucratic fear and rejection of the new technique. Almost universally, human cloning was condemned.[2-6] Germany, Denmark, and Spain have
legislation banning cloning; Norway, Slovakia, Sweden, and Switzerland have legislation implicitly banning cloning.[7] Some states in Australia, such as Victoria, ban
cloning. There are two bills before Congress in the United States which would comprehensively ban it.[8-9] There is no explicit or implicit ban on cloning in England,
Greece, Ireland, or the Netherlands, though in England the Human Embryology
and Fertilization Authority, which issues licenses for the use of embryos, has indicated that it would not issue any license for research into "reproductive cloning."

Reprinted with permission of BMJ Publishing Group from *Journal of Medical Ethics* 25
(1999): 87–95.

This is understood to be cloning to produce a fetus or live birth. Research into cloning in the first fourteen days of life might be possible in England.[7]

There have been several arguments given against human reproductive cloning:

1. It is liable to abuse.
2. It violates a person's right to individuality, autonomy, selfhood, etc.
3. It violates a person's right to genetic individuality (whatever that is—identical twins cannot have such a right).
4. It allows eugenic selection.
5. It uses people as a means.
6. Clones are worse off in terms of well-being, especially psychological well-being.
7. There are safety concerns, especially an increased risk of serious genetic malformation, cancer, or shortened lifespan.

There are, however, a number of arguments in favor of human reproductive cloning. These include:

1. General liberty justifications.
2. Freedom to make personal reproductive choices.
3. Freedom of scientific enquiry.
4. Achieving a sense of immortality.
5. Eugenic selection (with or without gene therapy/enhancement).
6. Social utility—cloning socially important people.
7. Treatment of infertility (with or without gene therapy/enhancement).
8. Replacement of a loved dead relative (with or without gene therapy/enhancement).
9. "Insurance"—freeze a split embryo in case something happens to the first: as a source of tissue or as replacement for the first.
10. Source of human cells or tissue.
11. Research into stem cell differentiation to provide an understanding of aging and oncogenesis.
12. Cloning to prevent a genetic disease.

The arguments against cloning have been critically examined elsewhere and I will not repeat them here.[10,11] Few people have given arguments in favor of it. Exceptions include arguments in favor of 7–12,[12] with some commentators favoring only 10–11[13,14] or 11–12.[15] Justifications 10–12 (and possibly 7) all regard cloning as a way of treating or avoiding disease. These have emerged as arguably the strongest justifications for cloning. This paper examines 10 and to some extent 11.

CLONING

HUMAN CLONING AS A SOURCE OF CELLS OR TISSUE

Cloning is the production of an identical or near-identical genetic copy.[16] Cloning can occur by fission or fusion. Fission is the division of a cell mass into two equal and identical parts, and the development of each into a separate but genetically identical or near-identical individual. This occurs in nature as identical twins.

Cloning by fusion involves taking the nucleus from one cell and transferring it to an egg which has had its nucleus removed. Placing the nucleus in the egg reprograms the DNA in the nucleus to replicate the whole individual from which the nucleus was derived: nuclear transfer. It differs from fission in that the offspring has only one genetic parent, whose genome is nearly identical to that of the offspring. In fission, the offspring, like the offspring of normal sexual reproduction, inherits half of its genetic material from each of two parents. Henceforth, by "cloning," I mean cloning by fusion.

Human cloning could be used in several ways to produce cells, tissues, or organs for the treatment of human disease.

Human Cloning as a Source of Multipotent Stem Cells

In this paper I will differentiate between totipotent and multipotent stem cells. Stem cells are cells which are early in developmental lineage and have the ability to differentiate into several different mature cell types. Totipotent stem cells are very immature stem cells with the potential to develop into any of the mature cell types in the adult (liver, lung, skin, blood, etc.). Multipotential stem cells are more mature stem cells with the potential to develop into different mature forms of a particular cell lineage, for example, bone marrow stem cells can form either white or red blood cells, but they cannot form liver cells.

Multipotential stem cells can be used as

a. a vector for gene therapy.
b. cells for transplantation, especially in bone marrow.

Attempts have been made to use embryonic stem cells from other animals as vectors for gene therapy and as universal transplantation cells in humans. Problems include limited differentiation and rejection. Somatic cells are differentiated cells of the body, and not sex cells which give rise to sperm and eggs. Cloning of somatic cells from a person who is intended as the recipient of cell therapy would provide a source of multipotential stem cells that are not rejected. These could also be vectors for gene therapy. A gene could be inserted into a somatic cell from the patient, followed by selection, nuclear transfer, and the culture of the appropriate clonal population of cells in vitro. These cells could then be returned to the patient as a source of new tissue (for example bone marrow in the case of leukemia) or as tissue without genetic abnormality (in the case of inherited genetic disease). The major experimental issues which would need to be addressed are developing clonal stability during cell amplification and ensuring differentiation into the cell type needed.[13] It should be noted that this procedure does not necessarily involve the production of a multicellular embryo, nor its implantation in vivo or artificially. (Indeed, cross-species cloning—fusing human cells with cow eggs—produces embryos which will not develop into fetuses, let alone viable offspring.[17])

A related procedure would produce totipotent stem cells which could differentiate into multipotent cells of a particular line or function, or even into a specific tissue. This is much closer to reproductive cloning. Embryonic stem cells from mice have been directed to differentiate into vascular endothelium, myocardial and skeletal tissue, hemopoietic precursors, and neurons.[18] However, it is not known whether the differentiation of human totipotent stem cells can be controlled in vitro. Unlike the previous application, the production of organs could involve reproductive cloning (the production of a totipotent cell which forms a blastomere), but then differentiates into a tissue after some days. Initially, however, all early embryonic cells are identical. Producing totipotent stem cells in this way is equivalent to the creation of an early embryo.

Production of Embryo/Fetus/Child/Adult as a Source of Tissue

An embryo, fetus, child, or adult could be produced by cloning, and solid organs or differentiated tissue could be extracted from it.

CLONING AS SOURCE OF ORGANS, TISSUE, AND CELLS FOR TRANSPLANTATION

The Need for More Organs and Tissues

Jeffrey Platts reports: "So great is the demand that as few as 5 percent of the organs needed in the United States ever become available."[19] According to David K. C. Cooper, this is getting worse: "The discrepancy between the number of potential recipients and donor organs is increasing by approximately 10–15 percent annually."[20] Increasing procurement of cadaveric organs may not be the solution. Anthony Dorling and colleagues write:

A study from Seattle, USA, in 1992 identified an annual maximum of only 7,000 brain-dead donors in the USA. Assuming 100 percent consent and suitability, these 14,000 potential kidney grafts would still not match the numbers of new patients commencing dialysis each year. The clear implication is that an alternative source of organs is needed.[21]

Not only is there a shortage of tissue or organs for those with organ failure, but there remain serious problems with the compatibility of tissue or organs used, requiring immunosuppressive therapy with serious side effects. Using cloned tissue would have enormous theoretical advantages, as it could be abundant and there is near perfect immunocompatibility.[22]

There are several ways human cloning could be used to address the shortfall of organs and tissues, and each raises different ethical concerns.

1. Production of Tissue or Cells Only By Controlling Differentiation

I will now give an argument to support the use of cloning to produce cells or tissues through control of cellular differentiation.

The fate of one's own tissue

Individuals have a strong interest or right in determining the fate of their own body parts, including their own cells and tissues, at least when this affects the length and quality of their own life. A right might be defended in terms of autonomy or property rights in body parts.

This right extends (under some circumstances) both to the proliferation of cells and to their transmutation into other cell types (which I will call the Principle of Tissue Transmutation).

Defending the Principle of Tissue Transmutation

Consider the following hypothetical example:

Lucas I. Lucas is a twenty-two-year-old man with leukemia. The only effective treatment will be a bone marrow transplant. There is no compatible donor. However, there is a drug which selects a healthy bone marrow cell and causes it to multiply. A doctor would be negligent if he or she did not employ such a drug for the treatment of Lucas's leukemia. Indeed, there is a moral imperative to develop such drugs if we can and use them. Colony-stimulating factors, which cause blood cells to multiply, are already used in the treatment of leukemia, and with stored marrow from those in remission in leukemia before use for reconstitution during relapse.

Lucas II. In this version of the example, the drug causes Lucas's healthy skin cells to turn into healthy bone marrow stem cells. There is no relevant moral difference between Lucas I and II. We should develop such drugs and doctors would be negligent if they did not use them.

If this is right, there is nothing problematic about cloning to produce cells or tissues for transplantation by controlling differentiation. All we would be doing is taking, say, a skin cell and turning on and off some components of the total genetic complement to cause the cell to divide as a bone marrow cell. We are causing a differentiated cell (skin cell) to turn directly into a multipotent stem cell (bone marrow stem cell).

Are there any objections? The major objection is one of practicality. It is going to be very difficult to cause a skin cell to turn *directly* into a bone marrow cell. There are also safety considerations. Because we are taking a cell which has already undergone many cell divisions during terminal differentiation to give a mature cell such as a skin cell, and accumulated mutations, there is a theoretical concern about an increased likelihood of malignancy in that clonal population. However, the donor cell in these cases is the same age as the recipient (exactly), and a shorter life span would not be expected. There may also be an advantage in some diseases, such as leukemia, to having a degree of incompatibility between donor and recipient bone marrow so as to enable the donor cells to recognize and destroy malignant recipient cells. This would not apply to non-malignant diseases in which bone marrow transplant is employed, such as the leukodystrophies. Most importantly, all these concerns need to be addressed by further research.

Lucas IIA. In practice, it is most likely that skin cells will not be able to be turned directly into bone marrow cells: there will need to be a stage of totipotency in between. The most likely way of producing cells to treat Lucas II is via the cloning route, where a skin cell nucleus is passed through an oocyte to give a totipotent cell. The production of a totipotent stem cell is the production of an embryo.

Production of an embryo as a source of cells or tissues

There are two ways in which an embryo could be a source of cells and tissues. Firstly, the early embryonic cells could be made to differentiate into cells of one tissue type, for example, bone marrow. Secondly, differentiated cells or tissues from an older embryo could be extracted and used directly.

CLONING

Are these permissible?

In England, the Royal Society[15] has given limited support to cloning for the purposes of treating human disease. The Human Genetics Advisory Commission (HGAC) defines this as "therapeutic cloning," differentiating it from "reproductive cloning."[7] Both bodies claim that embryo experimentation in the first fourteen days is permitted by English law, and question whether cloning in this period would raise any new ethical issues.

Cloning in this circumstance raises few ethical issues. What is produced, at least in the first few days of division after a totipotent cell has been produced from an adult skin cell, is just a skin cell from that person with an altered gene expression profile (some genes turned on and some turned off). In one way, it is just an existing skin cell behaving differently from normal skin cells, perhaps more like a malignant skin cell. The significant processes are ones of *cellular multiplication* and later, *cellular differentiation.*

If this is true, why stop at research at fourteen days? Consider the third version of the Lucas case:

Lucas III. The same as Lucas IIA, but in this case, Lucas also needs a kidney transplant. Therefore, in addition to the skin cell developing blood stem cells (via the embryo), the process is adjusted so that a kidney is produced.

The production of another tissue type or organ does not raise any new relevant ethical consideration. Indeed, if Lucas did not need the kidney, it could be used for someone else who required a kidney (if, of course, in vitro maturation techniques had been developed to the extent that a functioning organ of sufficient size could be produced).

Consider now:

Lucas IV. In addition to the blood cells, all the tissue of a normal human embryo is produced, organized in the anatomical arrangement of an embryo. This (in principle) might or might not involve development in a womb. For simplicity, let us assume that this occurs in vitro (though this is impossible at present).

Is there any morally relevant difference from the previous versions? It is not relevant that many different tissues are produced rather than one. Nor is the size of these tissues or their arrangement morally relevant. If there is a difference, it must be that a special kind of tissue has been produced, or that some special relationship develops between existing tissues, and that a morally significant entity then exists. When does this special point in embryonic development occur?

The most plausible point is some point during the development of the brain. There are two main candidates:

1. when tissue differentiates and the first identifiable brain structures come into existence as the neural plate around day 19.[23]
2. when the brain supports some morally significant function: consciousness or self-consciousness or rational self-consciousness. The earliest of these, consciousness, does not occur until well into fetal development.

On the first view, utilization of cloning techniques in the first two weeks to study cellular differentiation is justifiable. The most defensible view, I believe, is that our continued existence only becomes morally relevant when we become self-conscious. (Of course, if a fetus can feel pain at some earlier point, but is not self-conscious, its existence is morally relevant in a different way: we ought not to inflict unnecessary pain on it, though it may be permissible to end its life painlessly.) On this view, we should use the drug to cause Lucas IV's skin cells to transmutate and remove bone marrow from these. What is going on in Lucas IV is no different, morally speaking, from cloning. If this is right, it is justifiable to extract differentiated tissues from young fetuses which have been cloned.

CONCEPTION AND POTENTIALITY

The other usual point in development which is taken to be morally significant is conception.[24] However, in the case of cloning, there is no conception. There is just a process of turning some switches in an already existing cell. Proponents of the persons-begin-to-exist-at-conception view might reply that cloning is like conception. An individual begins to exist at the point of nuclear transfer. But why should we accept this? Conception seems quite different. Conception involves the unification of two different entities, the sperm and the egg, to form a new entity, the totipotent stem cell. In the case of cloning, there is identity between the cell before and after nuclear transfer—it is the same cell. Something new and important does happen to the entity when it undergoes nuclear transfer, just as something new and important happens when a cell with a malignant potential becomes malignant. But it is the same cell.

CLONING

Potentiality

In response, one might claim that after nuclear transfer the cell undergoes a radically and morally significant change: it acquires the potential to be a person. On this view, the cell immediately prior to nuclear transfer does not have the potential to be a human being but after nuclear transfer, it has the potential to be a human being. This has a jarring ring to it. What happens when a skin cell turns into a totipotent stem cell is that a few of its genetic switches are turned on and others turned off. To say it doesn't have the potential to be a human being until its nucleus is placed in the egg cytoplasm is like saying my car does not have the potential to get me from Melbourne to Sydney unless the key is turned in the ignition. (Rather, we should say that it has the potential but that that potential may not be *realized* if the key is not turned in the ignition.) Or it is like saying that a stick of dynamite *acquires* the potential to cause an explosion when placed in the vicinity of a lighted match. Of course, a stick of dynamite has the potential to cause an explosion, and various conditions, including placing it in the vicinity of a lighted match, are sufficient to realize this potential. In general terms, X has the potential to be Y, if X would be a Y if conditions c, d, e,…obtained. Nuclear transfer is like a number of other conditions (such as adequate placental blood flow) which must obtain if a skin cell is to become a person.

There may be another difference between a mature skin cell and a fertilized egg. Totipotent cells directly give rise to human beings but mature skin cells do not. The latter must go through a further stage of totipotency first. And it may be that that change is significant enough to say that the skin cell does not itself have the potential to create a human being. However, something with the potential to cause A may not lead directly to A. Killing the president may not lead directly to a world war. However, it may lead to political destabilization which will cause a world war. Killing the president does then have the potential to cause a world war.

At bottom, these issues may be semantic, and depend on how we choose to define "potential." What matters morally is whether skin cells *can* become human beings with the application of technology, and whether they *should*. That is an important moral feature of nuclear transfer. Nuclear transfer is a technical intervention which it is necessary to employ if a skin cell is to become a person, just as microsurgical transfer of an embryo formed in vitro is necessary if the embryo is to become a person.

I cannot see any intrinsic morally significant difference between a mature skin cell, the totipotent stem cell derived from it, and a fertilized egg. They are all cells which could give rise to a person if certain conditions obtained. (Thus, to claim that experimentation on cloned embryos is acceptable, but the same experimentation on noncloned embryos is not acceptable, because the former are not embryos but totipotent stem cells, is sophistry.)

Looking at cloning this way exposes new difficulties for those who appeal to the potential of embryos to become persons and the moral significance of conception as a basis for opposition to abortion. If all our cells could be persons, then we cannot appeal to the fact that an embryo could be a person to justify the special treatment we give it. Cloning forces us to abandon the old arguments supporting special treatment of fertilized eggs.

Production of a Fetus

If one believes that the morally significant event in development is something related to consciousness, then extracting tissue or organs from a cloned fetus up until that point at which the morally relevant event occurs is acceptable. Indeed, in law, a legal persona does not come into existence until birth. At least in Australia and England, abortion is permissible throughout fetal development.

Production of a Child or Adult as a Source of Cells or Tissues

Like the production of a self-conscious fetus, the production of a cloned child or adult is liable to all the usual cloning objections, together with the severe limitations on the ways in which tissue can be taken from donors for transplantation.

Many writers support cloning for the purposes of studying cellular differentiation because they argue that cloning does not raise serious new issues above those raised by embryo experimentation.[15] Such support for cloning is too limited. On one view, there is no relevant difference between early embryo research and later embryo/early fetal research. Indeed, the latter stand more chance of providing viable tissue for transplantation, at least in the near future. While producing a cloned live child as a source of tissue for transplantation would raise new and important issues, producing embryos and early fetuses as a source of tissue for transplantation may be morally obligatory.

CONSISTENCY

Is this a significant deviation from existing practice?

CLONING

1. Fetal Tissue Transplantation

In fact, fetal tissue has been widely used in medicine. Human fetal thymus transplantation is standard therapy for thymic aplasia or Di George's syndrome. It has also been used in conjunction with fetal liver for the treatment of subacute combined immunodeficiency.

Human fetal liver and umbilical cord blood have been used as a source of hematopoietic cells in the treatment of acute leukemia and aplastic anemia. Liver has also been used for radiation accidents and storage disorders. The main problem has been immune rejection.[25]

One woman with aplastic anemia received fetal liver from her own twenty-two-week fetus subsequent to elective abortion over twenty years ago.[26]

Fetal brain tissue from aborted fetuses has been used as source of tissue for the treatment of Parkinson's disease. Neural grafts show long-term survival and function in patients with Parkinson's disease, though significant problems remain.[27,28]

Fetal tissue holds promise as treatment for Huntington's disease,[29,30] spinal cord injuries,[31] demyelinating disorders,[27] retinal degeneration in retinitis pigmentosa,[32,33] hippocampal lesions associated with temporal lobe epilepsy, cerebral ischaemia, stroke and head injury,[34] and beta thalassemia in utero using fetal liver.[35] Fetal pancreas has also been used in the treatment of diabetes.

FETAL TISSUE BANKS

Indeed, in the United States and England, fetal tissue banks exist to distribute fetal tissues from abortion clinics for the purposes of medical research and treatment. In the United States, the Central Laboratory for Human Embryology in Washington, the National Diseases Research Interchange, and the International Institute for the Advancement of Medicine and the National Abortion Federation, all distribute fetal tissue.

In the UK, the Medical Research Council's fetal tissue bank was established in 1957 and disperses about 5,000 tissues a year.

2. Conception of a Noncloned Child as a Source of Bone Marrow: Ayala Case

Not only has fetal tissue been used for the treatment of human disease, but human individuals have been deliberately conceived as a source of tissue for transplantation. In the widely discussed Ayala case, a seventeen-year-old girl, Anissa, had leukemia. No donor had been found in two years. Her father had his vasectomy reversed with the intention of having another child to serve as a bone marrow donor. There was a one in four chance the child would be compatible with Anissa. The child, Marissa, was born and was a compatible donor and a successful transplant was performed.[36]

A report four years later noted: "Marissa is now a healthy four-year-old, and, by all accounts, as loved and cherished a child as her parents said she would be. The marrow transplant was a success, and Anissa is now a married, leukemia-free, bank clerk."[37]

Assisted reproduction (IVF) has been used to produce children to serve as bone marrow donors.[38] It is worth noting that had cloning been available, there would have been a 100 percent chance of perfect tissue compatibility and a live child need not have been produced.

OBJECTIONS

While there are some precedents for the proposal to use cloning to produce tissue for transplantation, what is distinctive about this proposal is that human tissue will be: (i) cloned and (ii) deliberately created with abortion in mind. This raises new objections.

Abortion Is Wrong

Burtchaell, a Catholic theologian, in considering the ethics of fetal tissue research, claims that abortion is morally wrong and that fetal tissue cannot be used for research because no one can give informed consent for its use and to use it would be complicity in wrongful killing.[39] He claims that mothers cannot consent: "The flaw in this claim [that mothers can consent] is that the tissue is from within her body but is the body of another, with distinct genotype, blood, gender, etc." Claims such as those of Burtchaell are more problematic in the case of cloning. If the embryo were cloned from the mother, it would be of the same genotype as her, and, arguably, one of her tissues. Now at some point a cloned tissue is no longer just a tissue from its clone: it exists as an individual in its own right and at some point has interests as other individuals do. But the latter point occurs, I believe, when the cloned individual becomes self-conscious. The presence or absence of a distinct genotype is irrelevant. We are not justified in treating an identical twin differently from a nonidentical twin because the latter has a distinct genotype.

In a society that permits abortion on demand, sometimes for little or no reason, it is hard to see how women can justifiably be prevented from aborting a fetus for the purpose of saving someone's life. And surely it is more respectful of the fetus, if the fetus is an object of respect, that its body parts be used for good rather than for no good purpose at all.

It Is Worse to Be a Clone

Some have argued that it is worse to be a clone.[40] This may be plausible in the sense that a person suffers in virtue of being a clone—living in the shadow of its "parent," feeling less like an individual, treated as a means and not an end, etc. Thus cloning in the Ayala case would raise some new (but I do not believe overwhelming) issues which need consideration. But cloning followed by abortion does not. I can't make any sense of the claim that it is worse to be a cloned cell or tissue. These are not the things we ascribe these kinds of interests to. Cloning is bad when it is bad for a person. Likewise, arguments regarding "instrumentalization" apply to persons, and not to tissues and cells.

Creating Life with the Intention of Ending It to Provide Tissue

Using cloning to produce embryos or fetuses as a source of tissue would involve deliberately creating life for the purposes of destroying it. It involves intentionally killing the fetus. This differs from abortion where women do not intend to become pregnant for the purpose of having an abortion.

Is it wrong deliberately to conceive a fetus for the sake of providing tissue? Most of the guidelines on the use of fetal tissue aim to stop women having children just to provide tissues.[41] The reason behind this is some background belief that abortion is itself wrong. These guidelines aim to avoid moral taint objections that we cannot benefit from wrong-doing. More importantly, there is a concern that promoting some good outcome from abortion would encourage abortion. However, in this case, abortion would not be encouraged because this is abortion in a very special context: it is abortion of a *cloned* fetus for medical purposes.

But is it wrong deliberately to use abortion to bring about some good outcome?

In some countries (for example those in the former Eastern bloc), abortion is or was the main available form of birth control. A woman who had intercourse knowing that she might fall pregnant, in which case she would have an abortion, would not necessarily be acting wrongly in such a country if the alternative was celibacy. When the only way to achieve some worthwhile end—sexual expression—is through abortion, it seems justifiable.

The question is: is the use of cloned fetal tissue the best way of increasing the pool of transplantable tissues and organs?[42]

An Objection to the Principle of Tissue Transmutation

Another objection to the proposal is that we do not have the right to determine the fate of all our cells. For example, we are limited in what we can do with our sex cells. However, we should only be constrained in using our own cells when that use puts others at risk. This is not so in transmutation until another individual with moral interests comes into existence.

Surrogacy Concerns

At least at present, later embryonic and fetal development can only occur inside a woman's uterus, so some of the proposals here would require a surrogate. I have assumed that any surrogate would be freely consenting. Concerns with surrogacy have been addressed elsewhere,[43] though cloning for this purpose would raise some different concerns. There would be no surrogacy concerns if the donor cell were derived from the mother (she would be carrying one of her own cells), from the mother's child (she would be carrying her child again), or if an artificial womb were ever developed.

Should We Give Greater Importance to Somatic Cells?

I have claimed that the totipotent cells of the early embryo, and indeed the embryo, do not have greater moral significance than adult skin cells (or indeed lung or colon or any nucleated cells). I have used this observation to downgrade the importance we attach to embryonic cells. However, it might be argued that we should upgrade the importance which we attach to somatic cells.

This is a reductio ad absurdum of the position which gives importance to the embryo, and indeed which gives weight to anatomical structure rather than function. If we should show special respect to all cells, surgeons should be attempting to excise the very minimum tissue (down to the last cell) necessary during operations. We should be doing research into preventing the neuronal loss which occurs normally during childhood. The desquamation of a skin cell should be as monumental, according to those who believe that abortion is killing persons, as the loss of a whole person. These claims are, I think, all absurd.

CLONING

Yuk Factor

Many people would find it shocking for a fetus to be created and then destroyed as a source of organs. But many people found artificial insemination abhorrent, IVF shocking, and the use of animal organs revolting. Watching an abortion is horrible. However, the fact that people find something repulsive does not settle whether it is wrong. The achievement in applied ethics, if there is one, of the last fifty years has been to get people to rise above their gut feelings and examine the reasons for a practice.

Permissive and Obstructive Ethics

Many people believe that ethicists should be merely moral watchdogs, barking when they see something going wrong. However, ethics may also be permissive. Thus ethics may require that we stop interfering, as was the case in the treatment of homosexuals. Ethics should not only be obstructive but constructive. To delay unnecessarily a good piece of research which will result in a lifesaving drug is to be responsible for some people's deaths. It is to act wrongly. This debate about cloning illustrates a possible permissive and constructive role for ethics.

CONCLUSION

The most justified use of human cloning is arguably to produce stem cells for the treatment of disease. I have argued that it is not only reasonable to produce embryos as a source of multipotent stem cells, but that it is morally required to produce embryos and early fetuses as a source of tissue for transplantation. This argument hinges on:

1. The claim that the moral status of the cloned embryo and early fetus is no different from that of the somatic cell from which they are derived.
2. The claim that there is no morally relevant difference between the fetus and the embryo until some critical point in brain development and function.
3. The fact that the practice is consistent with existing practices of fetal tissue transplantation and conceiving humans as a source of tissue for transplantation (the Ayala case).
4. An argument from beneficence. This practice would achieve much good.
5. An argument from autonomy. This was the principle of tissue transmuta-

tion: that we should be able to determine the fate of our own cells, including whether they change into other cell types.

This proposal avoids all the usual objections to cloning. The major concerns are practicality and safety. This requires further study.

The HGAC and the Royal Society have broached the possibility of producing clones for up to fourteen days: "therapeutic cloning." Those bodies believe that it is acceptable to produce and destroy an embryo but not a fetus. Women abort fetuses up to twenty weeks and later. We could make it mandatory that women have abortions earlier (with rapid pregnancy testing). However, we do not. Moreover, while the decision for most women to have an abortion is a momentous and considered one, in practice, we allow women to abort fetuses regardless of their reasons, indeed occasionally for no or bad reasons. If a woman could abort a fetus because she wanted a child with a certain horoscope sign, surely a woman should be able to abort a fetus to save a person's life.

I have been discussing cloning for the purposes of saving people's lives or drastically improving their quality. While we beat our breasts about human dignity and the rights of cells of different sorts, people are dying of leukemia and kidney disease. If a woman wants to carry a clone of her or someone else's child to save a life, it may not be society's place to interfere.

The recent development of human totipotent stem cell lines from embryonic tissue[44,45] means that we are closer to understanding cellular development and differentiation, generating the hope that we may be able to produce tissue for transplantation directly from totipotent stem cells without going to the stage of producing a mature embryo or fetus. But that is still some way off, and at present requires deriving the cell lines from embryonic tissue. The use of nuclear transfer may still be the best way to produce highly *compatible* tissue, even coupled with this technology.

We could address the shortage of tissue for transplantation now. We could routinely employ embryo-splitting during IVF and create embryo banks as a source of fetal tissue. Indeed, rather than destroying millions of spare embryos, we could use them as a source of human tissue. As opposed to using nuclear transfer as a source of tissue, such proposals could not be instituted with the consent of a person who both needs the tissue and is the source of the tissue. That is, we could not appeal to the Principle of Tissue Transmutation to justify these proposals, though they may be justifiable on other grounds.

CLONING

Acknowledgment

Many thanks to Jeff McMahan, Bob Williamson, David McCarthy, Edgar Dahl, Peter Singer, Lynn Gillam, and Ainsley Newson.

Funding

The Murdoch Institute and the Cooperative Research Centre for Discovery of Genes for Common Human Diseases.

Notes

1. I use the word "embryo" to cover all stages of development from the single-cell stage to the stage at which there is significant organogenesis (about the eighth week), after which I will refer to the growing human as a "fetus." I do not differentiate between a preembryo and an embryo.

2. National Bioethics Advisory Commission, *Cloning Human Beings* (Maryland: National Bioethics Advisory Commission, 1997).

3. World Health Organization, *Proposed International Guidelines on Ethical Issues in Medical Genetics and Genetics Services* (Geneva: WHO, 1998).

4. The European Parliament, *Resolution on Cloning.* Motion dated March 11, 1997. Passed March 13, 1997 (The European Parliament, 1997).

5. UNESCO, *Declaration on the Human Genome and Human Rights,* adopted on November 11, 1997(13), article 11 (UNESCO, 1997).

6. D. Butler, "Europe Brings First Ban," *Nature* 391 (1998): 219.

7. HGAC, HGAC Papers, *Cloning Issues in Reproduction, Science and Medicine.* Issued for comment January 1998 at http://www.dti.gov.uk/hgac/papers/papers_c.htm.

8. Human Cloning Prohibition Act of 1998, 105th Congress, 2d sess., S 1599

9. Human Cloning Prohibition Act of 1998, 105th Congress, 1st sess., HR 923.

10. J. Harris, "Good-bye Dolly? The Ethics of Human Cloning," *Journal of Medical Ethics* 23 (1997): 353–60.

11. P. Singer (forthcoming).

12. J. F. Childress, "The Challenges of Public Ethics: Reflections on the NBAC's Report," *Hastings Center Report* 27 (1997): 5.

13. A. Trounson, "Cloning: Potential Benefits for Human Medicine," *Medical Journal of Australia* 167 (1997): 568–69.

14. J. P. Kassirer and N. A. Rosenthal, "Should Human Cloning Research Be Off Limits?" *New England Journal of Medicine* 338 (1998): 905–906.

15. Council of the Royal Society, *Whither Cloning?* (London: The Royal Society: The UK Academy of Science, 1998), pp. 1–8.

16. Every cell division results in genetic differences between progeny which are not completely repaired by DNA repair mechanism. Though the vast majority of these differences have no functional implications, they are differences none the less and cellular division does not normally produce an exactly identical copy of DNA (Pannos Iannou, personal communication).

17. P. Cohen, "Organs without Donors," *New Scientist* (July 11, 1998): 4.

18. M. J. Weiss and S. H. Orkin, "In Vitro Differentiation of Murine Embryonic Stem Cells: New Approaches to Old Problems," *Journal of Clinical Investigation* 97 (1996): 591–95.

19. J. L. Platts, "New Directions for Organ Transplantation," *Nature* 392 (1998): 11–17.

20. D. K. C. Cooper, "Xenotransplantation—State of the Art," *Frontiers of Bioscience* 1 (1996): 248–65.

21. A. Dorling et al., "Clinical Xenotransplantation of Solid Organs," *Lancet* 349 (1997): 867–71.

22. It is not clear whether different patterns of X inactivation or the different mitochondrial complement of cloned cells will affect immunocompatibility.

23. Prior to this point, cells have the potential to develop into several tissue types and there is a lack of coordinated functional activity. For example, Michael Lockwood has argued that early on it is not determined which cells will become the fetus and which the placenta (M. Lockwood, "Human Identity and the Primitive Streak," *Hastings Center Report* [January–February 1995]: 45). It is not clear until much later which cells will become the brain and which the liver. Indeed, early on, the primitive embryo has the capacity to divide into identical twins. Lockwood's argument is that it is only when the definite precursor of a person's brain is formed that a discrete and determined individual can be said to exist.

24. Even an embryo formed by conception in the usual way may not become a person—a significant proportion are spontaneously aborted.

25. D. E. Vawter, W. Kearney, K. G. Gervaise, A. L. Caplan, D. Garry, and C. Tauer, *The Use of Human Fetal Tissue: Scientific, Ethical and Policy Concerns* (Minnesota: University of Minnesota, 1990), pp. 38–41.

26. E. Keleman, "Recovery from Chronic Idiopathic Bone Marrow Aplasia of a Young Mother after Intravenous Injection of Unprocessed Cells from the Liver (and Yolk Sac) of Her 22mm CR-length Embryo: A Preliminary Report," *Scandinavian Journal of Hematology* 10 (1973): 305–308.

27. O. Lindvall, "Neural Transplantation," *Cell Transplantation* 4, no. 4 (1995): 393–400.

28. T. B. Freeman, "From Transplants to Gene Therapy for Parkinson's Disease," *Experimental Neurology* 144, no. 1 (1997): 47–50.

29. K. M. Shannon and J. H. Kordower, "Neural Transplantation for Huntington Disease: Experimental Rationale and Recommendations for Clinical Trials," *Cell Transplantation* 5, no. 2 (1996):339–52.

30. T. B. Freeman, P. R. Sanberg, and O. Isacson, "Development of the Human Striatum: Implications for Fetal Striatal Transplantation in the Treatment of Huntington Disease," *Cell Transplantation* 4, no. 6 (1995): 539–45.

31. E. A. Zompa, L. D. Cain, A. W. Everhart, et al., "Transplant Therapy: Recovery of Function after Spinal Cord Injury," *Journal of Neurotrauma* 14, no. 8 (1997): 479–506.

32. M. del Cerro, E. S. Lazar, and D. Diloreto Jr., "The First Decade of Continuous Progress in Retinal Transplantation," *Microscopy Research and Technique* 36, no. 20 (1997): 130–41.

33. T. M. Litchfield, S. J. Whitely, and R. D. Lund, "Transplantation of Retinal Pigment Epithelial, Photoreceptor and Other Cells as Treatment for Retinal Degeneration," *Experimental Eye Research* 64, no. 5 (1997): 655–66.

34. A. K. Shetty and D. A. Turner, "Development of Fetal Hippocampal Grafts in Intact and Lesioned Hippocampus," *Progress in Neurobiology* 50, nos. 5–6 (1996): 597–653.

35. J. L. Touraine, "In Utero Transplantation of Fetal Liver Stem Cells into Human Fetuses," *Journal of Hematotherapy* 5, no. 2 (1996): 195–99.

36. J. Rachels, "When Philosophers Shoot from the Hip," *Bioethics* 5 (1991): 66–71.

37. Note in *Hastings Center Report* (May/June 1994): 2.

38. R. Lamperd, "Race for Life," *Sun Herald*, July 2, 1998, p. 1.

39. J. T. Burthchaell, "University Policy on Experimental Use of Aborted Fetal Tissue," *IRB: A Review of Human Subjects Research* 10 (1988): 7–11.

40. S. Holm, "A Life in the Shadow: One Reason Why We Should Not Clone Humans," *Cambridge Health Care Quarterly* 7 (1998): 160–62.

41. See reference 25: the whole book.

42. Some argue that opt-out systems of organ procurement, umbilical cord blood, and xenotransplantation offer viable alternatives. I would take an organ derived from one of my own cells, if concerns about aging and increased mutation were adequately addressed.

43. J. Oakley, "Altruistic Surrogacy and Informed Consent," *Bioethics* 6, no. 4 (1992): 269–87.

44. J. A. Thomson, J. Itskovitz-Eldor, S. S. Shapiro, M. A. Waknitz, J. J. Swiergiel, et al., "Embryonic Stem Cell Lines Derived from Human Blastocysts," *Science* 282 (1998): 1145–47.

45. M. J. Shamblott, J. Axelman, S. Wang, E. M. Bugg, J. W. Littlefield, et al., "Derivation of Pluripotent Stem Cells from Cultured Human Primordial Germ Cells," *Proceedings of the National Academy of Sciences* 95 (1998): 13, 726–31.

Human Reproductive Cloning Is Unethical Because It Undermines Autonomy
Commentary on Savulescu

Robert Williamson

I believe that there is a personal right, ethically based, to individuality, autonomy, and identity. This right correlates with the right not to be seen primarily as a "means," but is more fundamental. Reproductive cloning crosses a significant boundary in removing the single most important feature of autonomy: the fact that each of us is genetically unique and individual. Our genetic identity is an essential part of this individuality, and it is our genetic differences that explain why societies which attempt to impose environmental conditioning to achieve uniformity have not succeeded.

Identical twins are always raised in objection to this. Of course, the fact that nature gives outcomes does not mean that these outcomes are necessarily ethical if reproduced deliberately by man (for instance, there are several genetic syndromes where an infant is born without arms, but it would be unethical to reproduce these using assisted reproduction). However, I believe that in the context of a discussion on cloning it is more relevant that at the moment of conception the

Reprinted by permission of BMJ Publishing Group from *Journal of Medical Ethics* 25 (1999): 96–97.

CLONING

embryo is a single genetically unique individual with a mix of (more or less) randomly assorted maternal and paternal genomes combined in an individual way. The separation of twins into two individuals occurs during later embryonic development. Twins are unique and autonomous in their genetics. Even if they closely resemble each other, they do not resemble anyone else.

To what extent are the attributes we normally associate with individuality— personality, intelligence, appearance—due to the unique genome of the individual? While it is not yet possible to make a definitive and quantitative statement about this, twin studies all show a very high concordance for behavioral traits between identical twins as compared to nonidentical twins.[1,2] Genotype also determines many aspects of interaction between an individual and the environment, and parental genotype may also be related to the environment that is provided for children. The interpretation of these data is not simple, but I think most observers would agree that genotype is a major determinant of behavior. The precise relationship between genetic uniqueness and autonomy deserves to be explored in greater detail elsewhere, since the two are not identical (genetic uniqueness is acquired at fertilization, while autonomy arises somewhat later, associated with brain development).

Suppose we examine cloning as an extreme form of human reproduction, and consider the cloning of 100,000 persons with an identical genotype because all are derived from nuclei of cells of one individual. Would it be ethical to clone 100,000 persons from one genome? Clearly, most people would answer that this is unethical. I assume that Savulescu,[3] Burley and Harris,[4] and their colleagues would also regard it as unethical. If so, why? Because, I suggest, people appreciate that genetic uniqueness (in each new generation) is a property of humankind, with a fundamental and inseparable relationship to autonomy. It is understood that the 100,000 genetically identical individuals cloned from one person's genome would be diminished in autonomy compared to their "outbred" neighbors. If this argument holds for the creation of 100,000 clones from one person, it follows that it is also unethical to derive a single clone from an individual, for the ethical arguments are precisely the same. (Incidentally, this also implies that "embryo-splitting" to create twins from a newly fertilized two-cell embryo raises far fewer ethical issues than intergenerational reproductive cloning.)

One of the difficulties in many areas of ethics as applied to human medical genetics is that agendas may be illustrated using situations of desperate people who do not have time on their side. There are several thousand rare single-gene disorders, and many of them are catastrophic. Children still die of leukemia and other diseases because there are no appropriate tissues for transplantation. In clinical medicine we are used to giving our all, even breaking rules (such as financial or hierarchical rules) when faced with a seriously ill or dying patient, particularly a child. This leads Julian Savulescu to sanction human reproductive cloning as a remedy that may (possibly) save the lives of a few of these very hard cases.[3] However, hard cases make bad ethics in the same way as they make bad law.

A totipotent stem cell from an existing individual cannot be redefined as a new individual because it does not possess its own unique genome. Otherwise every totipotent cell, such as an embryonic stem cell, cultured from a human becomes "a new individual," which is clearly nonsense. It may soon be possible to modify cells from humans in culture so they lose their differentiated nature ("de-differentiate") and acquire the ability to re-differentiate to form other tissues. This would be completely justified in the absence of reproductive cloning. Lucas I and Lucas II have nothing to worry about! Nor does Lucas III, since the procedure does not lead to the generation of an individual capable of independent consciousness.

Lucas IV, on the other hand, has moved to a point where such an individual is in existence. This individual is not in existence before three weeks after fertilization (the development of the primitive brain), and is in existence at, say, thirty weeks (when most fetuses are capable of survival outside the womb). At some time between those two dates a step has been taken which crosses what I regard as a fundamental ethical barrier, the issue of genetic uniqueness and autonomy. We could argue whether it is closer to three weeks or thirty, but the principle appears to me to be clear—if one believes that there is an ethical value in the concept of genetic uniqueness at conception, then this barrier should not be breached other than for a more compelling imperative.

Julian Savulescu argues his cases well; we all are touched when confronted with the child dying of leukemia for want of a compatible donor, or the young adult in renal failure needing a kidney to be freed of the pain and hazard of renal dialysis. The long-term solution to these problems is to find a way of using "reproductive cloning technology" to provide cells, or even tissues, from the individual (or another compatible donor) which can be used effectively and in the numbers required without going through the reproductive cloning step. There would be nothing unethical in this. It will soon be possible for bone marrow transplantation. It may take a further short period of time to acquire the ability to grow tissues from de-differentiated cells from an individual in culture, perhaps a decade. Such a timescale does not appear to me to be unreasonable, nor such as to require us to revise our concepts of ethics in relation to autonomy by supporting reproductive cloning.

NOTES

1. C. C. Mann, "Behavioral Genetics in Transition," *Science* 264 (1994): 1686–89.

2. R. Plomin, M. J. Owen, and P. McGuffin, "The Genetic Basis of Complex Human Behaviors," *Science* 264 (1994): 1733–39.

3. J. Savulescu, "Should We Clone Human Beings? Cloning as a Source of Tissue for Transplantation," *Journal of Medical Ethics* 25 (1999): 87–95.

4. J. Burley and J. Harris, "Human Cloning and Child Welfare," *Journal of Medical Ethics* 25 (1999): 108–13.

RELIGION:
OFFICIAL STATEMENTS

eligious leaders and theologians have been in the thick of the debate about cloning. In this section, we present the responses of a number of Christian churches to the Dolly event. One thing that you will learn is that, just as philosophers have trouble over the correct interpretation and application of their moral principles—remember the debates over the Kantian Categorical Imperative—so also Christians have trouble over the correct interpretation and application of their moral principles. Love your neighbor as yourself, but what is love and who is your neighbor? Are you loving your neighbor as yourself if you clone them and produce offprints? Are you loving your neighbor as yourself if you clone them and harvest the embryo for spare parts?

Three of the denominations represented—the Church of Scotland (Presbyterians), the Catholics, and a branch of the Baptists—are all against human cloning in its entirety—against it in no uncertain terms. The Church of Scotland is willing to endorse animal cloning for medical reasons, but on theological grounds deplores a general application. It thinks that such a practice goes against God's creation, reducing or denying the variety which is a testament to God's greatness and intentions. In the case of humans, the prohibition is to be absolute. Again the dignity and uniqueness arguments make their appearance (suggesting apart from anything else that secular and religious moral philosophies are not necessarily all that different). Note that appeal is made to the Categorical Imperative, and that cloning is judged "unacceptable human abuse" because there is use of "the one cloned as a means to an end, for someone else's benefit." (Kant incidentally was the child of devote pietistic parents, so he would certainly not feel uncomfortable were he to find his ideas used by today's Christians.)

Catholics generally take a dim view of all attempts to change or interfere with or otherwise regulate the usual patterns of reproduction. Drawing on the thought of Saint Thomas Aquinas, they argue that the world is as it is because it was made by God and it is for us to follow and live within the limits set by this world. To do otherwise is to contravene "natural law." Artificial methods of fertilization, for instance, are immoral because they go against the natural order of events. Therefore, they condemn all attempts to clone humans, and given that they believe that a soul is implanted at the moment of conception and so all prebirth terminations are judged the moral equivalent of murder, they are also against any attempts to use embryos for harvesting or like purposes.

The Baptists are also against human cloning—they affirm "the sanctity and uniqueness of every human life." When we turn to the Methodists, however, although they call for a human cloning ban now, they seem less definitive in their opposition. Note that in respects their problems seem more consequentialist than Kantian—"we do not know all the consequences of cloning (psychological, social, or genetic)"—and they seem to envision the possibility that worries might be overcome and human cloning allowed. Note also that in some respect they seem to recognize that outright theological opposition could lead one into unfortunate moral absolutism. If one insists that a person born through cloning is necessarily thereby diminished, one is in some sense downgrading that person as a worthwhile human being. But if one is to avoid this trap, one has to allow that origins cannot be as all important as some insist.

Finally we have the United Church of Christ, one of the more liberal American Protestant denominations. They, too, call for a ban on human cloning for reproductive purposes—at least they call for such a ban at the moment, although they seem open to its possibility in the future. Note that although, later in their discussion of animal cloning, they make derogatory remarks about utilitarianism, here their own reasoning seems utilitarian. Their concern is that cloning puts too much human effort into the few and not enough into the masses. How can one justify cloning for the privileged, when "the world groans with hunger"? Note that like the Methodists, this group is also concerned that one not condemn a person—including a cloned person—simply on the grounds of their origin.

This church group however differs from other Christians in this section in that, even now, they would allow some cloning for research and therapeutic ends. They affirm the dignity of human life but they refuse to accord full human life to the early embryo. For this reason, they are prepared to accept some work at this stage, although they stress that it must be done with respect for the subject and must be accompanied by full discussion and inquiry. One must work within "a context of public understanding and general public support." Expectedly, when it comes to other organisms, including mammals, they have little general objection to research involving cloning. But liberal though it may be, the United Church of Christ is still Christian, and like the Presbyterians acknowledges the rich "God-given complexity of creation." For this reason, it too would regret and oppose the general introduction of cloning as a tool in agriculture and elsewhere.

Denominational Statements on Cloning

21

E ditor's note: These statements were available at the time this book went to press. Since then, other church bodies have also issued statements.

1. The General Assembly of the Church of Scotland,
   May 22, 1997

2. The Secretariat for Pro-Life Activities of the
   National Conference of Catholic Bishops.
   February 25, 1997

3. The United Methodist Genetic Science Task Force,
   May 9, 1997

4. The Christian Life Commission of the Southern Baptist Convention, March 6, 1997

5. The Genetics Committee of the United Church Board for Homeland Ministries of the United Church of Christ, June 9, 1997

"MOTIONS ON CLONING" PASSED BY THE GENERAL ASSEMBLY OF THE CHURCH OF SCOTLAND ON MAY 22, 1997

1. Commend the principle of the production of proteins of therapeutic value in the milk of genetically modified sheep and other farm animals but oppose, and urge Her Majesty's Government to take necessary steps to prevent, the application of animal cloning as a routine procedure in meat and milk production, as an unacceptable commodification of animals.

2. Reaffirm their belief in the basic dignity and uniqueness of each human being under God. Express the strongest possible opposition to the cloning of human beings and urge Her Majesty's Government to press for a comprehensive international treaty to ban it worldwide.

"CLONING ANIMALS AND HUMANS: A SUPPLEMENTARY REPORT TO THE 1997 GENERAL ASSEMBLY" FROM THE SOCIETY, RELIGION AND TECHNOLOGY PROJECT, BOARD OF NATIONAL MISSION

Introduction

In February 1997, Dolly the cloned sheep became a global news sensation. Scientists at the Roslin Institute and PPL Therapeutics outside Edinburgh had rewritten the laws of biology in producing a live sheep by cloning from cells of the udder of an adult ewe by nuclear transfer.

The Director of the Church of Scotland's Society, Religion and Technology Project quickly became a focus for ethical comment, with numerous TV, radio,

CLONING

newspaper and magazine interviews and articles, including BBC's *Newsnight* and *Heart of the Matter* programs, and German and Dutch national television. These opportunities arose because the SRT Project was already involved in assessing the issue, having for the last three years had an expert working group on the ethical issues of genetic engineering, one of whose members is the Roslin scientist who "produced" Dolly, Dr. Ian Wilmut. As a result, SRT was in a unique position to offer informed and balanced comment to the world's media, and to influence the course of a debate in which sensation and ill-judged speculation were rife.

A very significant development was the fact that quite a lot of media picked up SRT's involvement from the Internet. This was because Dr. Bruce had written an article on the ethics of cloning at the time of the first Roslin cloning discovery a year before, and had included this in SRT's own site on the World Wide Web. When the news broke, press agencies like CNN News searched for "cloning" on the Internet and found SRT's article as one of very few in existence. They put in a link to SRT from their own site, and all the world has since been following the trail—a month later, the SRT article was still receiving 400 Internet "visits" every day. This speaks volumes for the importance of SRT's work at the cutting edge of some of the most important issues which science is raising for our times. SRT identified four main issues—the basic genetic engineering work at Roslin, whether we should already clone animals, whether we might one day clone humans, and how such research should be controlled and kept accountable to the public.

Cloning Animals

Much of the media attention focused on speculation about cloning humans, but missed the more immediate impact on animals, and the ethical questions on how far we should apply technology to them. The cloning arose from Roslin's search for more effective ways to do the existing work at PPL Therapeutics of genetically engineering sheep to produce therapeutic proteins in sheep's and cow's milk. The first product for emphysema and cystic fibrosis sufferers is undergoing clinical trials, and a range of other medical applications is in prospect. The SRT Project working group had already found this work generally ethically acceptable. There were clear human benefits, with few animal welfare or other concerns once past the experimental stage.

The new method should enable Roslin to do a more precise genetic modification using less experimental animals, but the side effect is that the resulting sheep is a clone, genetically almost identical to its founder. PPL might clone 5–10 sheep from a single genetically modified animal, but would then breed naturally thereafter to give flocks of varied sheep, but all containing the desired genetic modification. On this very limited scale, this would not seem ethically unacceptable.

The possibility that farm animals might be cloned routinely for meat or milk production on a large scale is however a very different matter. In animal breeding, the need to maintain genetic diversity sets practical limits on how far cloning would make sense, but certain applications are already being considered. A breeder might wish to clone the best breeding stock to sell for feeding up for slaughter, or to found new nucleus herds. Would this be carrying our use of animals one stage too far?

For the Christian, the world around us is God's creation. Variety is one of its characteristic features, and especially at the level of higher animals and humans. The overall picture in the Bible, in commandments, stories, and poetry, is of a creation whose sheer diversity is itself a cause of praise to its creator. To reduce this diversity to a strict blueprint, and produce replica animals routinely on demand, would seem to go against something basic and God-given about the nature of life. The very fact that selective breeding has its limits reflects this fact. Some would argue that cloning is thus absolutely wrong, no matter what it was being used for. SRT argues that scale and intention play a part. PPL's limited context could be acceptable since the main intention was not the clone as such but growing an animal of a known genetic composition, where natural methods would not work. What would be unacceptable would be in routine animal production, where natural methods exist, but would be sidestepped on the grounds of economics or convenience. This would represent one step too far beyond conventional selective breeding in the way we use animals as commodities. The approach that, whatever use we find for animals, we could clone them to do it more efficiently brings the mass-production principles of the factory too far into the animal kingdom. Just as in the Old Testament an ox was not to be muzzled while treading out the grain, animals have certain freedoms which we should preserve. We may use animals to an extent, but we need to remind ourselves that they are firstly God's creatures, to whom we may not do everything we like.

Human Cloning

One of the abiding SciFi nightmares has been the idea that we could one day replicate human beings asexually, just by copying material from human cells. Roslin's scientists told a Select Committee of the House of Commons that the nuclear transfer technique they have applied to produce Dolly could be in theory applied to humans, but the headlines which ran "human cloning in two years" were irresponsible exaggerations. It is by no means a foregone conclusion. Dr. Wilmut and his colleagues made it quite clear that they think that to clone humans would be unethical, and most people seem to agree.

The Church of Scotland has already stated that to clone human beings would be ethically unacceptable as a matter of principle. On principle, to replicate any human technologically is a violation of the basic dignity and uniqueness of each human being made in God's image, of what God has given to that individual and to no one else. It is not the same as twinning. There is a world of difference ethically between choosing to clone from a known existing individual and the unpredictable occurrence of twins of unknown nature in the womb. The nature of cloning is that of an instrumental use of both the clone and the one cloned as means to an end, for someone else's benefit. This represents unacceptable human abuse, and a potential for exploitation which should be outlawed worldwide.

In 1990 the UK pioneered legislation making human cloning research illegal, but currently it would be allowed in the USA and several EU countries, and many other cultures with very different value systems. Some form of international treaty should be called for whereby no country would allow cloning research to be carried over from animals to humans. Realistically, there would be no way to stop a backstreet clinic or a dictatorship from ignoring such a treaty, but the lines need to be drawn. A second line of defense is also called for—the notion of the ethical scientist, for whom it would be against all professional principles to pursue such research. Some have argued that research should be permitted into the possibility of cloning living transplant organs from body cells. This would require more careful ethical consideration, but the danger of a "slippery slope" to full human cloning would loom large over such an enterprise.

This raises a last question of the control of such research. In many spheres of research there is a deficit in public accountability in the existing procedures whereby research priorities are set. There are no easy solutions to this problem, but, at the very least, it points to the need for a standing ethical commission on nonhuman biotechnology, whose work is open to public comment and scrutiny, in which those areas of research which are especially likely to have far-reaching ethical implications are first debated in public.

"REMARKS IN RESPONSE TO NEWS REPORTS ON THE CLONING OF MAMMALS," BY THE SECRETARIAT FOR PRO-LIFE ACTIVITIES OF THE NATIONAL CONFERENCE OF CATHOLIC BISHOPS, FEBRUARY 25, 1997

Recent reports about successful cloning in mammals have rightly raised ethical concerns about human cloning. Catholic teaching rejects the cloning of human beings, because this is not a worthy way to bring a human being into the world.

Children have a right to have real parents, and to be conceived as the fruit of marital love between husband and wife. They are not products we can manufacture to our specifications. Least of all should they be produced as deliberate "copies" of other people to ensure that they have certain desired features.

Donum Vitae, issued in 1987 by the Holy See's Congregation for the Doctrine of the Faith, reminded us that human life is a "gift of inestimable value" over which we must exercise careful stewardship. "The one conceived must be the fruit of parents' love," not treated as the product of a laboratory technique. Efforts to clone human embryos are also unethical because they would subject developing members of the human family, who cannot give informed consent, to risky experiments that cannot benefit them as individuals.

Such technologies should prompt us once again to appreciate a basic truth: The fact that it is technically possible to do something doesn't mean it ought to be done.

Richard M. Doerflinger
Secretariat for Pro-Life Activities
National Conference of Catholic Bishops

February 25, 1997

"STATEMENT FROM THE UNITED METHODIST GENETIC SCIENCE TASK FORCE," MAY 9, 1997

(This is a statement by the Genetic Science Task Force of the United Methodist Church. Only the General Conference speaks for the entire church. The Task Force was commissioned by the United Methodist General Board of Church and Society.)

Cloning has sparked enormous and sustained concern in the general public, including the church. It touches on many crucial questions about human nature, raises hopes and expectations, and brings to the fore uncertainties and fears. These include fears with substantial social and theological ramifications—fears that people will be used or abused, that women will be exploited, that the fabric of the family will be torn, that human distinctiveness will be compromised, fears that genetic diversity will be lessened, that corporate profit and personal gain will control the direction of research and development, and that privacy will be invaded. While we do not see obvious benefits of human cloning and recognize potential dangers of animal and tissue cloning, we also acknowledge the excitement that this new research generates for advances in medicine, agriculture, and other scientific endeavors.

As United Methodists, our reflections on these issues emerge from our faith. When we think about cloning, we remember that creation has its origin, value, and destiny in God, that humans are stewards of creation, and that technology has brought both great benefit and harm to creation. As people of faith, we believe that our identity as humans is more than our genetic inheritance, our social environment, or the sum of the two. We are created by God and have been redeemed by Jesus Christ. In light of these theological claims and other questions, fears, and expectations, we recognize that our present human knowledge on this issue is incomplete and finite. We do not know all the consequences of cloning (psychological, social, or genetic). It is important that the limits of human knowledge be considered as policy is made.

1. At this time, we call for a ban on human cloning. This would include all intended projects, privately or governmentally funded, to advance human cloning. (For the purposes of this document, human cloning means the intentional production of identical humans and human embryos.)

2. We call for a ban on therapeutic, medical, and research procedures which generate waste embryos.

3. As Christians, we affirm that all human beings, regardless of the method of reproduction, are children of God and bear the Image of God. If humans were ever cloned, they along with all other human beings, would have inherent value, dignity, and moral status and should have the same civil rights (in keeping with principles found in the United Nations Declaration of Human Rights). There must be no discrimination against any person because of reproductive origins. The same principles of autonomy, consent, and equality of opportunity would apply as fully to cloned human beings as to any other.

4. If research on human cloning does proceed, we urge that the impact on affected populations be given particular consideration—especially those whose voices often go unheard. These would include the women whose health and well-being might be affected, their offspring, the families, infertile men and women, and those who would be denied access to these technologies. We urge consideration of the psychological and social effects on individuals, families, parental relationships, and the larger society. Those presently affected by in vitro fertilization, surrogacy, artificial insemination, and other reproductive technologies should be consulted to provide insights into some related psychological and social issues.

5. We urge the widespread discussion of issues related to cloning in public forums including churches. Given the profound theological and moral implications, the imperfection of human knowledge, and the tremendous potential risks

and benefits, we urge that any moves toward cloning proceed slowly until these issues can be discussed fully both by the general public as well as by experts in agricultural and biological science, social science, public policy, ethics, theology, law, and medicine, including genetics and genetic counseling. In addition, discussion and policy making must include input from communities of faith.

6. Acknowledging that market forces play a significant role in the development of new technologies, we want appropriate social and governmental bodies to guide and monitor research and development in this field. Concern for profit and commercial advantage should be balanced by consideration for individual rights, the valid interest of wide constituencies, and the common good, including the good of future generations.

Questions to Be Addressed

Many questions must be addressed in a discussion of cloning, including the following:

Is there a compelling argument for human cloning? Are there enough clear benefits to outweigh the risks?

Is human cloning the only or best way to arrive at these benefits?

Is human cloning in the best interest of those affected—particularly children, women, families, and parents?

How do we exercise stewardship in the allocation of scarce resources? Would money and human effort be better spent on other forms of research?

How can societies effectively implement, enforce, and maintain cloning guidelines and regulations nationally and globally? What measures can be taken to discourage the misuse of these technologies and the exploitation of human beings?

"RESOLUTION ON CLONING" ADOPTED BY THE TRUSTEES OF THE CHRISTIAN LIFE COMMISSION OF THE SOUTHERN BAPTIST CONVENTION, MARCH 6, 1997

Whereas, on February 27, 1997, Ian Wilmut and colleagues in Scotland announced the first successful cloning of a mammalian species, and

Whereas researchers in Oregon announced March 1, 1997, the successful cloning of a rhesus monkey, and

Whereas on March 4, 1997, President Clinton announced a prohibition on federal funding for human cloning research, and

CLONING

Whereas Southern Baptists are on record for their strong affirmation of the sanctity and uniqueness of every human life;

Be it therefore resolved that we, the trustees of the Southern Baptist Convention's Christian Life Commission, meeting on March 6, 1997, do affirm the president's decision to prohibit federal funding for human cloning research, and

Be it further resolved that we request that the Congress of the United States make human cloning unlawful, and

Be it further resolved that we call for all nations of the world to make efforts to prevent the cloning of any human being.

"STATEMENT ON CLONING" BY THE UNITED CHURCH OF CHRIST COMMITTEE ON GENETICS

The following is a statement of the United Church of Christ Committee on Genetics and is not an official statement of the United Church of Christ. The Committee on Genetics was appointed by the United Church Board for Homeland Ministries, in cooperation with the United Church Board for World Ministries, the Office for Church in Society, and the Council for Health and Human Service Ministries. The committee will submit this statement for consideration as a resolution at the 21st General Synod of the United Church of Christ, meeting in Columbus, Ohio, July 3–8, 1997.

The announcement in February of the birth of Dolly the cloned sheep raises important ethical and religious questions. We who serve on the United Church of Christ Committee on Genetics are grateful for the careful and dedicated work of scientists such as Ian Wilmut and for the ways in which their discoveries can improve human health. We acknowledge that scientists themselves have been among the first to call for restraint in this field of research. We also applaud the request by President Clinton that the National Bioethics Advisory Commission (NBAC) prepare a report on the ethics of cloning, and that the NBAC eagerly sought the opinions of religious leaders in preparing its report. Because we believe that all new technologies should be developed with due regard for morality and social justice, we urge the continuation of the Bioethics Commission as a national forum for discussion of the ethics of emerging fields of research, including human cloning and germ-line experimentation.

1. Regarding the Use of Cloning to Produce a Human Person

The United Church of Christ is fundamentally committed to justice. We experience a tension with regard to cloning, therefore. On the one hand, we are aware that our culture has allowed, even encouraged, the development of many technologies geared to permit couples to have children "of their own"—meaning children who are genetically related to them. Cloning might allow some couples to have children whose genes come from that couple and are not in admixture with genes from "outside" the committed relationship. Thus, on the one hand, justice seems to press for allowing the same privileges (some would say rights) for those couples as are currently allowed for others.

At the same time, a concern for justice raises questions about the validity of all new reproductive technologies—artificial insemination, in vitro fertilization, etc. When the world groans with hunger, when children are stunted from chronic malnutrition, when people die of famine by the thousands every day—when this is the reality of the world in which we live, the development of any more technologies to suit the desires of those who are relatively privileged, secure, and comfortable seems to fly in the face of fundamental claims of justice.

For this reason, in spite of our empathy with couples who might seek cloning in order to have children "of their own," we oppose cloning and say "enough" to technologies that are privileges of the rich in the Western world. We support legislation to ban cloning for reproductive purposes, at least for the foreseeable future.

At the same time, should someone produce a cloned child, we would insist absolutely that such a child is a full human being, created in the image of God and entitled to all human and civil rights. This conviction is grounded both in our theological belief in the uniqueness of each person as a child of God and in the insights of genetics, which show us that genes and environment interact at every stage to produce the phenotype or the organism. Accordingly, we believe that while genes can be cloned or copied, the phenotype or the person cannot be cloned or copied but is always a unique expression of the interplay between genes and environment. We recognize that there is widespread confusion among the general public about the role of genes in determining who we are as persons. In this respect, genetic science and Christian faith agree in criticizing what is sometimes called "genetic determinism." We hope the United Church of Christ will play a helpful role in educating the public about the important but quite limited role of genes in determining the full meaning of human identity and personhood.

In addition to these fundamental concerns regarding justice and respect for all created life, members of the committee raised the following concerns:

First, there is evidence that the current state of technique of nuclear transfer cloning is far too imprecise to meet minimal expectations of safety that should be met before they are applied to human beings. Whether the safety level can be improved sufficiently through research with other mammals will remain to be seen. It is possible the future research will remove this concern. Our second and third reasons, however, are moral in nature and are not likely to be affected by the future level of technical ability.

Second, it is possible that a child produced by cloning would suffer from an overwhelming burden of expectations. Anyone wanting to create such a child would have a prior knowledge of what this child's genes could become and would in part make the decision to clone on the basis of that knowledge. That prior knowledge would create a weight of expectation against which such a child would have to define his or her own identity. Precisely because "genetic determinism" is so widespread in contemporary culture, this weight of expectation would likely be inconsistent with the freedom necessary for each person to develop an individual identity.

Third, many observers believe that it is beneficial for children to have the genetic resources of two adults that are recombined to form a genotype that is unique and yet tied genetically to both adults. This assures that, in terms of nuclear DNA, the child is related to both adults yet different from either. If children were produced by nuclear transfer cloning, their nuclear DNA would not have this relatedness and this difference.

2. Regarding the Use of Cloning to Produce Human Preembryos for Research

Nuclear transfer cloning might also be used to produce human preembryos for research purposes through the fourteenth day of embryonic development. It is very likely that through such means, scientists could learn a great deal about basic human developmental biology and that this knowledge might someday lead to treatments for degenerative conditions or to counteract some forms of sterility. Whether such research is permissible ultimately raises the question of the theological and moral status of the human preembryo. Beginning with the 8th General Synod in 1971, various General Synods of the United Church of Christ have regarded the human preembryo as due great respect, consistent with its potential to develop into full human personhood. General Synods have not, however, regarded the preembryo as the equivalent of a person.

Therefore, we on the United Church of Christ Committee on Genetics do not object categorically to human preembryo research, including research that produces and studies cloned human preembryos through the fourteenth day of fetal development, provided the research is well justified in terms of its objectives, that the research protocols show proper respect for the preembryos, and that they not be implanted. We urge public discussion of current research and future possibilities, ranging from preimplantation genetic screening of human preembryos to nuclear transfer cloning to human germ-line experimentation. We do not categorically oppose any of these areas of research, but we believe they must be pursued, if at all, within the framework of broad public discussion. In 1989, the 17th General Synod of the United Church of Christ stated that it was "cautious at present about procedures that would make genetic changes which humans would transmit to their offspring (germline therapy).... We urge extensive public discussion and, as appropriate, the development of federal guidelines during the period when germline therapy becomes feasible." We would urge legislation to ensure that all research on human preembryos, even that which is privately funded, would be reviewed by Institutional Review Boards in accordance with federal regulations.

Now with the possibility of human cloning by nuclear transfer, this call for public discussion becomes all the more urgent. Current U.S. federal law prohibits the use of federal funding for all preembryo research. Such research occurs in other countries and in the United States, funded by private sources. As a result, human preembryo research proceeds legally in the United States, but there is little or no public discussion of its ethics.

We on the United Church of Christ Committee on Genetics are opposed to the idea that human preembryo research, such as human germination experimentation or research involving cloned preembryos, should be permitted but left largely unregulated if funded privately, or that there is no federal responsibility for the ethics of such research if federal funds are not used. We believe that this approach merely seeks to avoid the difficult public deliberation that should occur prior to such research. We believe that all such research should be subject to broad public comment and that it should only proceed within a context of public understanding and general public support.

3. *Regarding Cloning or Nonhuman Mammalian Species*

We on the United Church of Christ Committee on Genetics believe that the use of nuclear transfer cloning in research on nonhuman mammalian species is morally and theologically permissible, provided, of course, that animals be treated humanely and that needless suffering is avoided. Nevertheless, we are concerned that the use of nuclear transfer cloning, together with other genetic and reproductive technologies, will contribute to a diminished regard for nonhuman species. In particular, we lament the attitude that nonhuman species have no

CLONING

inherent dignity or significance beyond their usefulness to human beings. We confess with regret that in the past, the Christian church itself has often encouraged such a utilitarian view of nonhuman species. In contrast to our own past, the General Synod of the United Church of Christ stated in 1989 that "God is Creator of all and confers value upon all creatures.... Therefore, we respect each creature as valuable to God beyond its apparent usefulness to us." Such an attitude of respect does not entirely preclude our use of animals in research or for other purposes. But we are concerned that cloning, together with other technologies, might contribute to the view that nonhuman animals, particularly mammals, are little more than pharmaceutical factories, or convenient sources of donor organs for human patients, or valuable research tools. While we do not object specifically to any of these uses of mammalian nuclear transfer cloning, we are concerned nonetheless that these uses will contribute to a disregard for the dignity that all nonhuman species enjoy by virtue of their relation to God the Creator. Further, because of the danger of narrowing genetic diversity (and thus of diminishing the God-given complexity of creation) we do not think that animal cloning should ever be widely used, for instance in agricultural application.

RELIGION: PERSPECTIVES

T his section deal with reactions to and commentary on the religious response to Dolly. Ronald Cole-Turner starts his discussion with a brief review of the official (and semi-official) statements that were given in the last section. Then he goes on to systematize and analyze the reasons behind these statements—reasons which, he admits, are not always as fully articulated as one might wish. Especially, Cole-Turner points out, those not that familiar with Christian theology might have trouble following the full context of the Church responses. To this end, he goes over some of the pertinent territory: human dignity and its various meanings in the theological context (there is a good discussion here of the "humans made in the image of God" claim as a ground of human dignity); the meaning and place of the family in Christian thought; and the question of justice in a Christian context. One point Cole-Turner brings out and emphasizes is the extent to which the Christian is committed to acceptance of God's ways and the hope of His grace in this scheme, rather than to the belief that we ourselves can solve all of life's problems through our own unaided efforts. This point—the belief that the world is providentially ordered rather than potentially ever-more improvable through human progress—is clearly highly relevant to discussions about a new technology like cloning. We have already seen a version of it in Kass's case against human cloning.

Joshua H. Lipschutz looks at cloning from a Jewish perspective. People who are not themselves Jews or much acquainted with Jewish thought and practice often think that there is something unworldly, old-fashioned, and restrictive about Judaism. Those food prohibitions and the silliness about not driving a car on the Sabbath! The more knowledgeable and more sympathetic realize that actually Judaism is a complex religion, with more to give than it takes. Particularly through its sacred and traditional writings, it offers much for the thinking religious man and woman who seeks guidance for proper living, even in a highly technological society such as ours. "The Jewish tradition offers an extremely rich history and a unique tradition of scholarly debate that has lasted for many millennia and covers an enormous range of topics." Against this background, Lipschutz finds no absolute moral or religious prohibition against human cloning with the aim of producing new persons. He finds it to be a natural process within the meaning of Jewish law and hence that a person born of such a process is to be considered fully human. Moreover, if a such a person is born of a Jewish woman—it is apparently the birth mother who counts, not the genetic mother—then he or she is automatically to be considered Jewish.

Finally, we have the astringent comments of Ronald A. Lindsay. He has little patience with the religious discussion of human cloning, the title of his piece—"Taboos without a Clue"—giving a fair indication of his conclusion. He finds the religious writers to be dogmatic without genuine support, self-justifying without external appeal, and offering "metaphor masquerading as reasoned argument." Having run through the various objections to human cloning made by people speaking in the name of religion, Lindsay stresses that he is not against moral discussion of cloning—not even against moral discussion of cloning by people with religious commitments—but rather that he is against such discussion other than on secular ethical grounds. We must appeal to reason and not to supposed supernatural revelation.

Cloning Humans from the Perspective of the Christian Churches

Ronald Cole-Turner

1. SCOPE OF THE LITERATURE

Pre-1997. There was an intense religious debate over cloning in the early 1970s, centered largely on the work of Paul Ramsey[1] and Joseph Fletcher.[2] In various church statements on genetics, beginning in the mid-1970s, cloning has been ignored.

1997–98. The February 1997 announcement of mammalian cloning prompted a flurry of statements by churches and religious leaders. Notable among church statements are those of the Church of Scotland (COS), the United Methodist Church (UMC), the Southern Baptist Convention (SBC), and the United Church of Christ (UCC). Various Roman Catholic documents are included, as is a press statement by the Christian Coalition.

Reprinted from *Science and Engineering Ethics* 5, no. 1 (1999): 33–46, published by Opragen Publications, Guildford, UK.

2. Topics Addressed

a. Animals. The question of the morality of animal cloning is generally over-looked in the religious discussion, with two exceptions:

The Church of Scotland:

> [We . . .] commend the principle of the production of proteins for therapeutic value in the milk of genetically modified sheep and other farm animals, but oppose, and urge Her Majesty's Government to take necessary steps to prevent the application of animal cloning as a routine procedure in meat and milk production, as an unacceptable commodification of animals (p. 138).[3]

The United Church of Christ:

> We on the United Church of Christ Committee on Genetics believe that the use of nuclear transfer cloning in research on nonhuman mammalian species is morally and theologically permissible, provided, of course, that animals be treated humanely and that needless suffering is avoided. Nevertheless, we are concerned that the use of nuclear transfer cloning, together with other genetic and reproductive technologies, will contribute to a diminished regard for non-human species. . . . [and] that cloning, together with other technologies, might contribute to the view that nonhuman animals, particularly mammals, are little more than pharmaceutical factories, or convenient sources of donor organs for human patients, or valuable research tools. While we do not object specifically to any of these uses of mammalian nuclear transfer cloning, we are concerned nonetheless that these uses will contribute to a disregard for the dignity that all nonhuman species enjoy by virtue of their relation to God the Creator. Further, because of the danger of narrowing genetic diversity (and thus of diminishing the God-given complexity of creation) we do not think that animal cloning should ever be widely used, for instance in agricultural application (p. 151).[4]

Note that while both statements support animal cloning research and pharmaceutical applications, they argue for a theocentric, as distinct from a purely anthropocentric, view of the worth and status of animals (Cf. Byers, p. 68).[5]

b. Nonreproductive Uses. Nonreproductive human cloning is overlooked in most of the religious discussion. For the Roman Catholic Church, however, the definitive teaching is set out in a 1987 document, *Instruction on Respect for Human Life*, which expressly prohibits any form of human embryo research that lacks a clear therapeutic intent.[6] The embryo possesses the full dignity of a human person, and thus any instrumental use of the embryo is forbidden. The Pontifical Academy statement on cloning says: "A prohibition of cloning which would be limited to preventing the birth of a cloned child, but which would still permit the cloning of an embryo-fetus, would involve experimentation on embryos and fetuses and would require their suppression from birth—a cruel, exploitative way of treating human beings. . . . Such experimentation is immoral."[7]

The UMC cloning document states: "We call for a ban on therapeutic, medical, and research procedures which generate waste embryos" (p. 144).[8] The Southern Baptist Convention Resolution stated: "we call on Congress to enact federal legislation against producing human embryos for the purpose of experimentation, whether by tax-funded or privately funded researchers."[9] The Christian Coalition, in a statement supporting legislation to ban experimental as well as reproductive human cloning, declares its "opposition to this clearly amoral and unethical science and we join today in calling for a total ban on cloning of human beings and human embryos. We do this consistent with our belief that human life begins at conception with a human embryo. . . ."[10]

By contrast, the UCC statement leaves the door open to human embryo research and possible medical applications:

> . . . we on the United Church of Christ Committee on Genetics do not object categorically to human preembryo research, including research that produces and studies cloned human preembryos through the fourteenth day of fetal development, provided the research is well justified in terms of its objectives, that the research protocols show proper respect for the preembryos, and that they not be implanted. We urge public discussion of current research and future possibilities, ranging from preimplantation genetic screening of human preembryos to nuclear transfer cloning to human germ-line experimentation. We do not categorically oppose any of these areas of research, but we believe they must be pursued, if at all, within the framework of broad public discussion (p. 150).[4]

An open but somewhat more cautious view is expressed by the Executive Committee of the European Ecumenical Commission for Church and Society:

> At this stage [June 1998], it would be premature to rule out a priori all such limited uses of cloning methods, but without knowing how they might be done, it is hard to assess their ethical validity. . . . [T]he creation of cloned human embryos for research purposes would seem unacceptable. It would be ridiculous to permit the creation of the beginnings of a potential individual which one knew in advance one would have to terminate, because its existence as a full human being would be considered ethically unacceptable.[11]

Compare this to the statement of moral theologian Gilbert Meilaender before the National Bioethics Advisory Committee (NBAC): "If we are genuinely baffled about how best to describe the moral status of that human subject who is the unimplanted embryo we should not go forward in a way that peculiarly combines metaphysical bewilderment with practical certitude by approving even such limited cloning for experimental purposes."[12] While Meilaender sets a higher threshold than the UCC statement for metaphysical or theological agreement before permitting research to proceed, both agree that research without public discussion and some level of agreement is imprudent for science and objectionable as public policy.

c. Reproductive Cloning. Most of the attention of the churches has been focused on the possibility of cloning as a human reproductive technique. Nearly all the churches and church leaders who have spoken have opposed this development. Their arguments are varied and often not clearly stated or fully developed. Quite often the arguments are grounded in theological warrants that are unstated or unclear to others. Furthermore, it is not always clear how far the conclusions of the arguments might be carried. Some arguments against reproductive cloning may be construed as arguments against all reproductive technology. While that may be intended by some churches or leaders, it is not fully intended by others. [I think it is fair to say that the prospect of reproductive cloning has prompted some to reconsider their support for other forms of reproductive technology.] Nor do all those who offer religious arguments against cloning agree with every other religious arguments against cloning. It is best to view these statements as appealing to a cluster of arguments with various levels of support by their proponents. At the same time, there is widespread agreement throughout the churches, I believe, that this cluster of objections deserves careful attention within the churches and in public, and that reproductive cloning should be banned at least until there has been sufficient time for serious discussion of moral and religious concerns.

In Sections 3–5, I will try to organize and explicate the core Christian objections to reproductive cloning. Some of these concerns will be shared by adherents to other faith traditions, even though they might object to the specifically Christian way in which they are stated here. It is fair to say, as does the NBAC report, that "religious perspectives on cloning humans differ in fundamental premises, modes of reasoning, and conclusions. As a result, there is no single 'religious' view on cloning. . . " (NBAC, 57).[13] Nevertheless, NBAC took religion very seriously and the report encourages continued public discussion of these concerns. In a secular society that is religiously pluralistic, however, it is unclear, to put it mildly, what role religious concerns should play in public debate, or how far religious citizens should permit their religion to define their citizenship.

3. Reproductive Cloning Would Be an Assault to Human Dignity

A. The Uses of "Human Dignity" as an Argument. The concern that cloning would violate human dignity is probably the most widely stated objection to cloning. It is found in religious statements of many traditions and in secular statements as well. The Church of Scotland statement warns that reproductive cloning would violate the "basic dignity" of human beings. The Catholic *Instruction on Respect for Human Life* states that all attempts at "obtaining a human being without any connection with sexuality through 'twin fusion,' cloning or parthenogenesis are to be considered contrary to the moral law, since they are in opposition to the dignity both of human procreation and of the conjugal union."[6] A statement of the Pontifical Academy, "Reflections on Cloning," bases its rejection of cloning on the ground that "it denies the dignity of the person subjected to cloning and the dignity of human procreation" (p. 5).[7] The Christian Coalition says that cloning disrespects "the decency and dignity of human life."[10]

We find similar words in secular documents. In the Universal Declaration on the Human Genome and Human Rights of UNESCO, Article 11, we read: "Practices which are contrary to human dignity, such as reproductive cloning of human beings, shall not be permitted."[14] And the French National Consultative Ethics Committee report states:

> There is therefore not a single conceivable variation of reproductive cloning of human beings, be it cloning of an adult or of an embryo, which is safe from an accumulation of intractable objections. For all of these reasons, it can only provoke vehement, categorical, and absolute ethical condemnation. Such a practice, which imperils radically the autonomy and dignity of the human person, would be a grave moral regression in the history of civilization. One might well consider whether the concept of degrading violation of the human condition, of which reproductive cloning is a clear example, should not be legally qualified with a view to a universal ban.[15]

We must note here, however, that the secular statements are confined to objections to reproductive cloning while the religious statements object to all forms of human cloning. So whatever is meant by "human dignity," those who use it to object to cloning on its basis seem divided on whether embryos possess it. Furthermore, the UCC statement does not raise any objection to cloning on the grounds of dignity: in fact the statement never uses the word "dignity" in reference to human beings but only to other animals. So it is quite obvious that the meaning of "human dignity" needs clarification. What is meant by human dignity, and how might cloning violate it?

B. Meanings of "Human Dignity." In these documents we can find at least three meanings of human dignity. Each has its limits and its problems as a basis for moral argument.

i. dignity as freedom. In statements by UNESCO or the French Ethics Committee, we find dignity linked to autonomy or freedom. Cloning is an assault to dignity because it is an infringement upon freedom, presumably because by human action the genome of another individual is determined, and this determination constrains the freedom of the clone. The Pontifical Academy statement worries "that some individuals can have total dominion over the existence of others, to the point of programming their biological identity—selected according to arbitrary or purely utilitarian criteria."[7] But we can readily imagine a strong counterargument: based on what we know about genetics, how would a clone be less free than its original? Psychological factors might come to bear here, but if they could be mitigated, the argument against cloning based on dignity as freedom would be difficult to sustain.

A more careful statement of the relationship between dignity and freedom is offered by the Executive Committee of the European Ecumenical Commission:

> To choose to replicate the genetic part of human make up technologically has been described as a violation of the basic dignity and uniqueness of each human being made in God's image, but it is important to establish more precisely what is at issue, the main questions center on the degree of control one human being has over the makeup of another, the instrumental way in which cloning would tend to use other human beings, and the risk of serious damage in psychological and relationship aspects. The biblical picture of humanity implies that we are far more than just our genes, but genes are a fundamental part of our makeup. By definition, to clone is to exercise unprecedented control over that dimension of another individual. Such control by one human over another is incompatible with the ethical notion of human freedom, in the sense that each individual's genetic identity is inherently unpredictable and unplanned.[11]

The concept of dignity as freedom, however, carries with it a peculiarly pro-cloning twist. A "Declaration in Defense of Cloning and the Integrity of Scientific Research," signed by many prominent scientists and scholars, makes this statement: "Such guideline [to prevent cloning abuses] should respect to the greatest extent possible the autonomy and choice of each individual human being. Every effort should be made not to block the freedom and integrity of scientific research."[16] The difficulty here is quite obvious. It is hard to see why we should prohibit cloning on the grounds that it violates human dignity, autonomy, and freedom; on the contrary, people should be free to request it and scientists should be free to investigate it.

ii. dignity as uniqueness. In the COS statement, we read:

> On principle, to replicate any human technologically is a violation of the basic dignity and uniqueness of each human being made in God's image, of what God has given to that individual to no one else. It is not the same as twinning. There is a world of difference ethically between choosing to clone from a known existing individual and the unpredictable occurrence of twins of unknown nature in the womb. The nature of cloning is that of an instrumental use of both the clone and the one cloned as means to an end, for someone else's benefit. This represents unacceptable human abuse, and a potential for exploitation which should be outlawed worldwide.[3]

Among religious writers, this argument has been subject to the most criticism (cf. Lebacqz,[17] Peters[18]). It makes one claim (uniqueness is inviolable) but in view of twinning switches to another claim ("instrumental use . . . as a means to an end" is offensive). It is not twinning but the technology of twinning that is offensive. If so, then the argument would best be stated as:

iii. dignity as resistance to instrumental use, objectification, and commodification. An interesting psychological form of this argument is offered in the French Ethics Committee statement: "Unlike Dolly, human clones would know they are clones and would know that others see them as clones. One cannot be blind to the intolerable lowering of a person to the status of an object that would ensue."[15] This argument is rarely developed explicitly in church statements, but it has been strongly developed in the statements of individual religious scholars. Some worry that in an era of reproductive liberty backed by numerous technologies including cloning, children will be seen as a technical accomplishment, as products instead of persons, used as an end (such as the fulfillment of parental desires), and that they will be accepted conditionally rather than loved unconditionally. For most of these Christian writers, these arguments are grounded in a theological understanding of human personhood and family, which in turn is rooted in an understanding of the nature of God, in whose image we find our true selves.

C. The "Image of God" as the Ground of Human Dignity

In Christian theology, the concept of dignity grows out of more fundamental concepts. Chief among these is that human beings are created in the image of God. In the arguments against cloning, we find references to this doctrine. Human dignity also grows out of the doctrine of redemption: human beings possess an inalienable dignity because God has granted them the grace of redemption, even to the point of taking up our humanity in the incarnation of Jesus Christ. The cloning documents scarcely mention this second ground for a theology of dignity, perhaps because it is exclusively Christian and emphatically not shared by those outside Christianity. Indeed, even "image of God" language may be suppressed in Christian documents for a similar reason: "dignity" seems more universal. But if "image of God" is edited down to "dignity," rich worlds of theological meaning are lost. Furthermore, the various and somewhat conflicting strands of meaning that make up the concept "image of God" are lost. Among other things, the "image of God" means:

i. the special capacities of human beings, notably reason and language. From this flows the idea that human beings occupy a place in nature that is superior to other forms of life. We are to be stewards of the creation, exercising a benign dominion over the earth and its creatures. Some have recently suggested that we are to play a role alongside God the Creator in the ongoing development of the creation. We are to use technology, including genetics, as a way to heal nature and enhance its potentialities. According to this view, human beings are God's "cocreators," and technology is a form of "cocreation." In this view, technology is not "playing God" but rather may be seen as a theologically legitimate form of participating with God in the process of creation. Philip Hefner uses the idea of "cocreation" in a statement that suggests that cloning may be fully acceptable: "We are created cocreators (some will say created by God, others, by nature): creatures of nature who themselves intentionally enter into the process of creating nature in startling ways. We face even the prospect of creating ourselves, in ways that are startling and troubling."[19]

While this last idea has not yet received explicit criticism, it should be noted that for most Christians, human beings are created in the image of God (and thus share in God's creative powers), but there are limits that God has put on our creativity. The biblical basis for these ideas is Genesis 1–11, which begins with humanity in the image of God but continues its narration with stories of betrayal, fratricide, an arms race, and an effort to use technology as a way to achieve union with the gods.

ii. that persons live in community. To be in the image of God is to be a person in relation. This view is grounded in the Christian understanding of God as Triune, existing as three Persons in such intimacy that the relationship itself is God. Because we are in God's image, our humanity is similarly communal: we are not isolated individuals who only coincidentally enter into contracts with others, but persons-in-community. Persons come to exist in community and they create other persons in community.

Contemporary theologians have recovered this view of the Trinity and its anthropological correlates. Indeed, one of the most common aspects of the contemporary Christian doctrine of God is its move toward relationality and its critique of individualism for God or for humanity. Even the NBAC report echoes this move by pointing to a key motif in the Christian idea of the Trinity, namely, that to be a person in union with other persons is to be "begotten and not made." According to Gilbert Meilaender, "begotten not made" is the ground of our equality with each other: "Although we are not God's equal we are of equal dignity with each other and we are not at each other's disposal. If it is, in fact, human begetting that expresses our equal dignity we should not lightly set it aside in a manner as decisive as cloning" (p. 63).[12] Children are an outgrowth of a loving relationship and not an act of will or of planning (technology). "The child is, therefore, always a gift. One like them [the parents] who springs from their embrace, not a being whom they have made and whose destiny they should determine" (p. 63).[12]

In several ways, this meaning of the image of God is at odds with the previous one. This one is communal where the previous one tends to be individualistic. Furthermore, this one does not look for intrinsic qualities in the individual as the condition for dignity but confers its status on the basis of relationship: a child is accepted as a child, as an outcome or expression of parental love, and not on the basis of rationality or other traits. "Children are not routinely interviewed and invited to join a family, nor are they acquired as an accoutrement adorning an already established relationship" (p. 83).[20] Human life is valued without qualification even when it is fragile, vulnerable, and genetically "imperfect." Our confidence lies not in our health or in our technology as a means to secure it in the promise that we are in relationship with God and with others. Nowhere is this outlook more tested than in the face of death and grief. Nevertheless, Christians will very likely refuse the idea of cloning a dead child:

> Familial love is vulnerable. An expansive and unfolding love is accompanied by unexpected joy and happiness, and also unanticipated sorrow, suffering, and loss. These qualities are inseparable if the family is to provide an unconditional place of belonging. Hence parenthood is always accompanied by the twin elements of hope and risk. Attempting to replace a dead child with a clone, say, is to turn familial love in upon itself, for it tries to reach back and recapture what has been lost. Yet an unfolding familial love embraces the pain and suffering of its members, making them a part of its ongoing and enfolding life. No member of a family can truly be replaced, for a family is not simply a structure built with

interchangeable parts, but a symbiotic affinity in its own unique right (pp. 85–86).[20]

iii. that our species develops individually and collectively. The meaning of the image of God, for Christians, is found not so much in the idea of creation as in the hope of redemption. We—individually and collectively—are moving towards the image of God. The content of this hope is defined in Jesus Christ, fully human in communion with other human beings, and yet intimately related to God. Our destiny is to develop toward maturity in Christ. "Progress" is judged entirely by the extent to which our lives and our society conform more fully to the "kingdom of God." Towards this end technology might play some role, as might human action in history. But the grounds for this transformation lie in divine grace and not in human achievement apart from that grace. Therefore any purely techno-logical quest for perfection is immediately suspect. Indeed, human action can thwart the intentions of God. Nevertheless, God is redemptive and so recalls us from our mistakes. Precisely because of this, Christian statements inevitably affirm what appears to be two conflicting ideas: cloning is an assault on human dig-nity, and a cloned person would be a person of fully equal dignity, for the clone, like everyone else, would possess dignity by virtue of a relationship which God establishes and which human beings cannot negate.

What about the soul of the clone? In the French Ethics Committee statement, we read: "It would even seem to be understood by some that an individual pro-duced by cloning is a kind of reincarnation of the clone's original."[15] The thought that cloning is a kind of reincarnation is of course not the Christian tradition and would never be defended in church statements. Indeed. "God does not send our souls back for another try. Any efforts to use cloning to 'try to bring someone back' would be scientifically and theologically misguided" (p. 125).[21]

The "Statement in Defense of Cloning" warns that "some religions teach... that humans have been imbued with immortal souls by a deity, giving them a value that cannot be compared to that of other living things. Human nature is held to be unique and sacred."[16] It is true that many traditions within Christianity hold to such a dualistic view of the human soul and body. But dualism was not a part of the earliest Christian view. It is being widely rejected today by contemporary the-ologians. Indeed, if we were dualists, we probably would not worry theologically about what someone might do to human genes or to the human brain. Precisely because we see the human person as a psychosomatic unity, we believe that genetic interventions, such as cloning, are theologically significant, affecting the core of the human person in all its relationships.

An interesting question, however, does arise on account of cloning for those

Christians who hold to a particular version of dualism. Some believe that God creates each individual soul at conception and joins it to the developing embryo. Now if by cloning an embryo comes into existence long after it is "conceived" when would God create its soul?

4. REPRODUCTIVE CLONING WOULD SUBVERT THE FAMILIAL CONTEXT OF HUMAN LIFE

Catholic documents have long charged that reproductive technology disrupts the meaning of the family as the context in which human life is engendered. In that respect the documents speak not only of the dignity of the person or the embryo but of "the dignity of human procreation." There should be little surprise, then, that the Catholic Church would not reject reproductive cloning. What may be surprising is that Protestant churches and leaders are concerned that cloning would violate the familial context of human life.

For instance, the UCC statement entertains (but does not fully endorse) this concern: "Many observers believe that it is beneficial for children to have the genetic resources of two adults that are recombined to form a genotype that is unique and yet tied genetically to both adults. This assures that in terms of nuclear DNA, the child is related to both adults yet different from either. If children were produced by nuclear transfer cloning, their nuclear DNA would not have this relatedness and this difference" (p. 149).[4] This concern is elaborated: "The cloning of humans should not be attempted because it disrupts the ordering of the family's natural and social affinity, distorts the family as a place of unconditional belonging, and violates the character of an unfolding and enfolding familial love" (p. 87).[20]

And:

> Is there an order of nature that would be violated by bringing a person into existence by cloning? The argument might run like this: The earliest organisms produced asexually, essentially by cloning. A momentous evolutionary advance occurred with sexual reproduction, allowing the reshuffling of genes in every generation and permitting far more genetic diversity to appear. Evolution could then occur more quickly and go in many directions, including our own. Furthermore, as creatures become more complex and need parents to rear them, sexual reproduction assured that both parents have a genetic interest in their offspring, so sexual reproduction can be seen as ultimately encouraging the care of offspring by two parents. For human beings, this parental investment must be sub-

stantial if the child is to thrive. Sexual reproduction creates a genetic bond between both parents and the child. At the same time, the child is unique—that is, a combination of both parents but a copy of neither. Cloning, by returning us to asexual reproduction, would violate this natural order that is so beneficial to the child. Other reproductive techniques, such as in vitro fertilization, if the sperm and egg come from the couple, respect the order of nature in that the child is a unique combination of the parents' genes (p. 127).[21]

In testimony before NBAC, a Catholic moral theologian stated the concern in similar ways:

> Up until now every human child has had two parents. The biological relationship between parents and children is a symbol of reproductive, social, and domestic partnership with great personal and social significance. Historically and cross-culturally families in all their variety of cultural form have been key institutions for the structuring of societies. A cloned individual will have a biogenetic link to one lineage only.... So the child who is truly the child of a single parent would be a genuine revolution in human history and her or his advent should be viewed with immense caution. In my view it is not too strong to say that cloning is a violation of the essential reality of human family (p. 51).[22]

Underlying this view is the conviction that God orders the natural and social world in ways that best conform to God's providential care, and that human life is best lived in consent to this order. The order that God intends for nature and for human society is not of our choosing and not subject to our revision.

Whether cloning (and only cloning) disregards that order is, of course, a matter for more study and debate. But it is not unreasonable to be concerned that cloning will confuse family relationships:

> When I come face to face with my clone, whom would I behold? This question is prompted not so much by confusion entailed in trying to determine whether this new being is my son, brother, or both, but rather, because, given this method of reproduction, these differentiated roles cannot emerge and develop.... I am not saying I could not love or care for my clone, I think I could. I simply do not know how our relationship could be defined within the context of a family. We would both be deprived of a full sense of familial relatedness and belonging. For it is the distinctive character of the spousal, parental, filial, and sibling relationships which shapes the ordering of a family's natural and social affinity (p. 84).[20]

5. REPRODUCTIVE CLONING WOULD CONTRIBUTE TO INJUSTICE RATHER THAN TO JUSTICE

For Christians, justice is not so much the defense of the status quo as it is a redemptive transformation of relationships towards respect, equality, and community. The end of justice is the "kingdom of God" on earth. Our task, individually and collectively, is to work toward that end. So when a new possibility such as human cloning arises, Christians will ask: Will it contribute to justice or detract from it?

Out of that concern, the UMC statement expresses its "fears that people will be used or abused, that women will be exploited, that the fabric of the family will be torn, that human distinctiveness will be compromised, fears that genetic diversity will be lessened, that corporate profit and personal gain will control the direction of research and development, and that privacy will be invaded" (p. 143).[8]

The UCC statement identifies justice as the primary theological concern regarding reproductive cloning:

> The United Church of Christ is fundamentally committed to justice. We experience a tension with regard to cloning, therefore. On the one hand, we are aware that our culture has allowed, even encouraged, the development of many technologies geared to permit couples to have children "of their own"—meaning, children who are genetically related to them. Cloning might allow some couples to have children whose genes come from that couple and are not an admixture with genes from "outside" the committed relationship. Thus, on the one hand, justice seems to press for allowing the same privileges (some would say rights) for those couples as are currently allowed for others.
>
> At the same time, a concern for justice raises questions about the validity of all new reproductive technologies—artificial insemination, in vitro fertilization, etc. When the world groans with hunger, when children are stunted from chronic malnutrition, when people die of famine by the thousands every day—when this is the reality of the world in which we live, the development of any more technologies to suit the desires of those who are relatively privileged, secure, and comfortable seems to fly in the face of fundamental claims of justice.
>
> For this reason, in spite of our empathy with couples who might seek cloning in order to have children "of their own," we oppose cloning and say "enough" to technologies that are privileges of the rich in the Western world. We support legislation to ban cloning for reproductive purposes, at least for the foreseeable future.[4]

Will cloning be good for children? It has been pointed out that of all the problems faced by the world's children, a shortage of good cloning is not one of them.

CLONING

What will cloning do to children? Will it leave other children worse off? Would the cloned child enter life deprived of any benefit? Would the cloned child begin life with an unusual and unjust burden, perhaps an overwhelming burden of expectation?

Will cloning be good for women? Will some women benefit while others do not? This discussion has been almost entirely neglected (important exception: Lebacqz[17]). The NBAC report has no section devoted to this theme.

Will cloning be good for racial minorities? At the very least, there is a plausible worry that cloning will lead ultimately to a severely narrowed view of what is normative humanity. Somewhat less likely is the thought that cloning might be used to create servants. These concerns are vivid and truly worrisome in light of recent eugenic projects in the United States, Sweden, and Germany, and in view of the experiment at Tuskegee (cf. Paris).[23]

Will cloning mean that we neglect other needs? When considered in light of the crushing needs of many in today's world, cloning and related technologies seem to be a selfish indulgence. The difficulty facing this argument, however, is that the development and use of cloning will likely be funded not by federal tax revenue but entirely by private funds. A large burden of proof awaits anyone who would argue that people cannot use their own money selfishly.

CONCLUSION

While all the church statements have expressed opposition to reproductive human cloning, there is no consensus among the churches or their leaders about any one of the objections to human cloning, nor is there likely to be. In part this is because there is little clarity about whether cloning differs from other human reproductive technologies, which enjoy support from some churches but encounter strong opposition for other churches. Even those churches which grant approval to reproductive technology generally are opposed to reproductive cloning.

There is some disagreement among the churches about the morality of experimental and therapeutic uses of human cloning. The strong and unequivocal opposition of some churches to embryo research certainly extends to cloning. But churches which do not share that unequivocal opposition are unclear about the theological and moral status of the embryo as a "potential person" and about the meaning of respect for such an entity.

There is agreement among the churches that religion bears substantially on public discussions about technology generally and about cloning in particular, and that the privatization and trivialization of religion should be resisted. In addition there is agreement that the various arguments against reproductive cloning deserve careful public discussion and that, at the very least, there should be a moratorium on reproductive cloning until such discussion occurs, probably for a period of ten to twenty years.

NOTES

1. P. Ramsey, *Fabricated Man: The Ethics of Genetic Control* (New Haven, Conn.: Yale University Press, 1970).

2. J. Fletcher, *The Ethics of Genetic Control: Ending Reproductive Roulette* (Garden City, N.Y.: Doubleday, 1974).

3. Church of Scotland, "Motions on Cloning" and "Supplementary Report," Edinburgh, reprinted in *Human Cloning: Religious Responses,* ed. R. Cole-Turner (Louisville, Ky.: Westminster John Knox Press, 1997), pp. 138–41.

4. United Church of Christ, "Statement of the Committee on Genetics," Cleveland, reprinted in Cole-Turner, *Human Cloning,* pp. 147–51.

5. D. Byers, "An Absence of Love," in Cole-Turner, *Human Cloning,* pp. 66–77.

6. "Congregation for the Doctrine of the Faith, Instruction on Respect for Human Life in Its Origin and on the Dignity of Procreation" (1987), reprinted in *Gift of Life: Catholic Scholars Respond to the Vatican Instruction,* ed. E. Pellegrino, J. C. Harvey, and J. P. Langan (Washington, D.C.: Georgetown University Press, 1990), pp. 1–41.

7. Pontifical Academy, "Reflections on Cloning" (October 1997), available at http://www.vatican.va.

8. United Methodist Church General Board of Church and Society, "Statement from the United Methodist Genetic Science Task Force" (May 9,1997), reprinted in Cole-Turner, *Human Cloning,* pp. 143–45.

9. Southern Baptist Convention, "Resolution on Genetic Technology and Cloning" (Dallas, Tex.: 1997).

10. Christian Coalition, "Statement by Heidi H. Stirrup, Director of Government Relations" (January 29, 1998).

11. Executive Committee of the European Ecumenical Commission for Church and Society, *Cloning Animals and Humans—An Ethical View* (Strasbourg: June 1998).

12. G. Meilaender, "Testimony to the National Bioethics Advisory Commission" (March 13, 1997) (Washington, D. C.: Eberlin Reporting Service), pp. 62–64. Reprinted as "Begetting and Cloning," *First Things* 74 (June/July 1997): 41–43.

13. National Bioethics Advisory Commission (United States), "Cloning Human Beings: Report and Recommendations" (June 9, 1997) (Rockville, Md.: NBAC).

14. UNESCO, "Universal Declaration on the Human Genome and Human Rights" (December 20, 1996), available at http://www.unesco.org/ibc/uk/genome/projet/index.html.

15. Comite Consultatif National D'Ethique pour les Sciences de la Vie et de la Santé (France), "Reply to the President of the French Republic on the Subject of Reproductive Cloning" (Report #54) (English translation) (Paris, France: The Comité, April 22, 1997), 26 pp. (Online) available at: http://www.ccne-ethique.org/ccne_uk/avia/a_054.htm.

16. "Statement in Defense of Cloning and the Integrity of Scientific Research" (May 20, 1997), *Chronicle of Higher Education* Web site: http://chronicle.com/che-data/focus.dir/data.dir/0520.97/cloning.htm.

17. K. Lebacqz, "Genes, Justice, and Clones," in Cole-Turner, *Human Cloning*, pp. 49–57.

18. T. Peters, "Cloning Shock: A Theological Reaction," in Cole-Turner, *Human Cloning*, 12–24.

19. P. Hefner, "Cloning as Quintessential Human Act," *Insights* (June 1997).

20. B. Waters, "One Flesh? Cloning, Procreation, and the Family," in Cole-Turner, *Human Cloning*, pp. 78–90.

21. R. Cole-Turner, "At the Beginning," in Cole-Turner, *Human Cloning*, pp. 119–30.

22. L. Cahill, "Testimony to the National Bioethics Advisory Committee" (March 13, 1997) (Washington, D.C.: Eberlin Reporting Service), p. 51.

23. P. J. Paris, "A View from the Underside," in Cole-Turner, *Human Cloning*, pp. 43–48.

To Clone or Not to Clone
A Jewish Perspective

Joshua H. Lipschutz

"A human being—born of clonal reproduction—most likely will appear on the earth in the next twenty to fifty years, and conceivably even sooner."[1]

James D Watson, Ph.D., Nobel Laureate

The headlines of February 23, 1997 said it all: "Scientist reports first ever cloning of adult mammal."[2] This was something that was supposed to be impossible because adult tissue, and by extension DNA, had always been assumed to be terminally and irreversibly differentiated.[3] There had been hints that the prevailing dogma was wrong, including work we had done demonstrating for the first time that an adult mammalian organ could be transformed into another organ in situ.[4] Still, few scientists believed that the cloning of an adult mammal was possible, let alone technically feasible. It should be noted that there is very little difference among mammals with respect to DNA; meaning the cloning of human beings now poses no significant technological hurdles. With this scientific bomb-

Reprinted with permission of BMJ Publishing Group from *Jounal of Medical Ethics* 25 (1999): 105–107.

shell[5] came a series of hasty declarations, including a United States presidential ethics commission recommendation to enact a limited ban on human cloning and a law declaring it illegal in California. The title of a very recent newspaper article: "On Cloning Humans, 'Never' Turns Swiftly into 'Why Not?'" demonstrates that the issue of human cloning is now being reexamined.[6] This paper asks the question: "Is cloning morally wrong?" and tries to answer it from a Jewish perspective.

One might argue that the issue of cloning is a highly technical, modern problem and an ancient religion would, therefore, be an inappropriate place to turn for help. However, the Jewish tradition offers an extremely rich history and a unique tradition of scholarly debate that has lasted for many millennia and covers an enormous range of topics. For example, regarding human artificial insemination, which didn't exist in ancient times, there is an obscure midrashic account of the birth of Ben Sira, the third-century author of the Proverbs of Ben Sira. The term "midrash" means investigation and its purpose is to explain the biblical text from an ethical point of view. This particular midrashic story describes how Ben Sira's mother became pregnant after immersion in a ritual bath where a leftover drop of sperm had fertilized her ovum. Because of the fact that Ben Sira is quoted in the Talmud, which is the transmission of the oral Torah to the people of Israel, it can be extrapolated that Ben Sira's origins were not enough to render him "un-Kosher" and that artificial insemination must therefore be permissible.

A logical place to begin the investigation of the ethical permissibility of human cloning is the Bible and one doesn't have very far to go to find the first clue. "And the man gave names to all cattle, and to the fowl of the air, and to every beast of the field; but for Adam there was not found a helpmeet for him. And the Lord G-d caused a deep sleep to fall upon the man, and he slept; and He took one of his ribs, and closed up the place with flesh instead thereof. And the rib, which the Lord G-d had taken from the man, made He a woman, and brought her unto the man."[7] This sounds suspiciously like cloning and one might reason that if G-d had done it, it must be okay. Still this isn't technically cloning because you could never make a woman from a man because of the difference in the two sex chromosomes (men are XY while women are XX) and it could be argued that precisely because G-d did it, it is forbidden (i.e., only G-d has the right to create life).

FABLES AND LEGENDS

In the Jewish tradition there are also many fables and legends that, while not literally true, are still felt to contain truths. One of the most famous is the four-hundred-year-old story of the Golem. The Golem, a powerful giant, was created out of clay and brought magically to life by Rabbi Judah Loew (the Maharal of Prague) to save his people from the violence and death that had resulted from the "blood lie," a preposterous rumor that Jews were using the blood of Christian children to make their Passover matza. The Golem, with brutal strength, single-handedly repelled the mob that had descended on the ghetto to wreak havoc. The creation of the Golem, though an act of man, was felt by most Jews to be a good thing and was not so different from how G-d had created the original Adam: "Then the Lord your G-d formed man of the dust of the ground, and breathed into his nostrils the breath of life; and man became a living soul." But what of the life that was created? After all, when the Golem had fulfilled his purpose and brought security to the Jews of Prague, Rabbi Loew undid his spell and returned the Golem to lifeless clay. Does this mean that the "life" that the Maharal of Prague created was any less holy or sacrosanct?

The answer to this can be found in the commentaries on the Mishnah. The Mishnah is the collection of Jewish law and ethics that ranks second in importance only to the Bible. The Mishnah forms the basis of the Talmud and is divided into six "parts"; each "part" is further subdivided into tractates. The fourth "part," called Neziken, contains a tractate called Sanhedrin, which deals with the courts of justice and judicial procedure, especially with reference to criminal law.[8] The eight-hundred-year-old commentary by Rabbi Shlomo Ha-Meiri explains that there is a difference between life created by magic, as in the case of the Golem, and life created by natural processes.[9] Cloning is certainly a natural process; therefore, any life created by cloning that is born of a mother must be a full life with all the rights and privileges accorded any human being. According to legend, the Golem created by the Maharal of Prague was actually not the first. The famous scholar Rava, who lived in the fourth century C.E. also created a Golem,[10] affording the rabbis an opportunity to address even the question of asexual reproduction. Further evidence for the correctness of Rabbi Ha-Meiri's interpretation is found in the very definition of who is a Jew. A full member of the Jewish community is any person born of a Jewish mother (or, of course, anyone who converts to Judaism).[11] Taking this idea one step further, there does not necessarily even have to be a father. Based on similar reasoning, the State of Israel in 1996 ruled that in the case of surrogate mothers, the birth mother rather than the genetic mother is the true mother. If the genetic parents want the child to be theirs, they have to adopt the child from the surrogate mother. Because a human being that was cloned would be born via a natural process of a human mother, there does not appear to be anything wrong, at least in the eyes of Halacha (Jewish Law), with cloning.

Regardless of the morality involved, one might still argue that it is not wise to have a number of "identical" people inhabiting this planet, that this would somehow diminish the individuality of man, which is something Western society holds sacred. An article by Johnson entitled "Don't Worry a Brain Can't Be Cloned"[12] goes a long way towards helping to resolve this criticism. The article begins by talking about the Aborigines and how they are terrified of cameras because of the fear that someone will steal their soul. To people in the "developed" world, the knowledge that a photograph is only skin deep, makes the Aborigines' fears seem absurd. However, the fear that one's very identity might be stolen, that one could cease to be an individual runs very deep even in "advanced" societies like ours, accounting for the success of movies like *The Net*, in which Sandra Bullock's identity is erased by a computer.

UNIQUENESS

Both the person that was cloned and the clonee would almost certainly wonder if they had lost their uniqueness, the very essence of what makes a human being special. For now, this is an abstract debate—but soon, I am convinced, this will be a very real question. The answer may be very simple. While the body may be cloned, the brain, which is the essence of humanity, will always be unique. Unlike other body tissues such as bone and muscle, which are made exactly according to a genetically predetermined plan, the brain and its neural networks grow and change with each experience. The precise layout of the neurons in the brain is what makes the difference. The neurons, which are linked to one another through junctions called synapses, form the circuitry that makes us who we are. During development, the genes lay out the basic wiring plan in the brain according to the instructions contained in the DNA. After the human being is born, and perhaps even before, the trillions of bits of sensory input that we call "experiences" make, break, and reform the billions of neuronal connections, creating a unique pattern that could never be duplicated because no two persons' experiences are ever the same. Even identical twins, which form when a fertilized egg splits in two, yielding two genetically identical beings, essentially nature's clones, have very different neuronal wiring and different likes and dislikes.[12]

As Johnson notes, you could keep two identical twins in the same room and their brains would still develop differently. One twin would go around the room clockwise while the other would go counterclockwise and their "experiences" would be different and their neural circuits would develop differently and they would develop into two separate beings. Even if someday man could do the ultimate cloning and copy, synapse by synapse, the human brain, the technological feat would be fleeting and the "identical" minds would only last for a brief instant. Suppose that neuron No. 20478288 were to fire randomly in brain one and not brain two. This tiny difference would set off a cascade that would reshape the circuitry and again there would be two individuals.[12] The human mind is beautiful and immeasurably complex and it is what gives us our uniqueness.

There are many circumstances where one could argue that human cloning would be desirable, such as where there is a need to clone someone to obtain bone marrow that could save a cancer patient's life or when there is a wish to clone the only child of a middle-aged couple who has died in a car accident. Of course there are dangers to cloning, even in the isolated cases mentioned above; however, this does not mean that cloning itself is wrong. Many new reproductive methods such as artificial insemination, in vitro fertilization, freezing of human embryos, and surrogate motherhood were first condemned but have come to be accepted. Perhaps the question should be changed from "Is cloning wrong?" to "When is cloning wrong?"

ACKNOWLEDGMENTS

The author would like to thank Emily Mostov for her very helpful discussion of the text.

This work was supported by NIH Grant DK02509 and a National Kidney Foundation Young Investigator Award.

NOTES

1. J. D. Watson, "The Future of Asexual Reproduction," *Intellectual Digest* (October 1971): 69–74.

2. G. Kolata, "Scientist Reports First Ever Cloning of Adult Mammal," *New York Times*, February 23, 1997, p. 1.

3. J. M. W. Slack, *From Egg to Embryo: Determinative Events in Early Development* (New York: Cambridge University Press, 1985).

4. J. H. Lipschutz, P. Young, O. Taguchi, and G. R. Cunha, "Urothelial Transformation into Functional Glandular Tissue In Situ by Instructive Mesenchymal Induction," *Kidney International* 49 (1996): 59–66.

5. I. Wilmut, A. E. Schnieke, J. McWhir, A. J. Kind, and K. H. S. Campbell, "Viable Offspring Derived from Fetal and Adult Mammalian Cells," *Nature* 385 (1997): 810–13.

CLONING

6. G. Kolata, "On Cloning Humans, 'Never' Turns Swiftly into 'Why Not?'" *New York Times*, December 2, 1997, p. 1.

7. The Bible, Gen. 2:20–22.

8. P. Birnbaum, *Jewish Concepts* (New York: Hebrew Publishing, 1975).

9. Sanhedrin 67:B. Commentaries on Talmud Bavli.

10. Sanhedrin 65:B. Talmud Bavli.

11. Kiddushin 68:B. Talmud Bavli.

12. G. Johnson, "Don't Worry a Brain Can't Be Cloned," *New York Times*, Week in Review section, March 2, 1997, p. 1.

Taboos without a Clue
Sizing Up Religious Objections to Cloning
Ronald A. Lindsay

SIZING UP RELIGIOUS OBJECTIONS TO CLONING

The furor following the announcement of recent experiments in cloning, including the cloning of the sheep Dolly, has prompted representatives of various religious groups to inform us of God's views on cloning. Thus, the Reverend Albert Moraczewski of the National Conference of Catholic Bishops has announced that cloning is "intrinsically morally wrong" as it is an attempt to "play God" and "exceed the limits of the delegated dominion given to the human race." Moreover, according to Reverend Moraczewski, cloning improperly robs people of their uniqueness. Dr. Abdulaziz Sachedina, an Islamic scholar at the University of Virginia, has declared that cloning would violate Islam's teachings about family heritage and eliminate the traditional role of fathers in creating children. Gilbert Meilender, a Protestant scholar at Valparaiso University in Indiana, has stated that cloning is wrong because the point of the clone's existence "would be grounded

Reprinted with permission from *Free Inquiry* 17, no. 3 (Summer 1997): 15–17.

in our will and desires" and cloning severs "the tie that united procreation with the sexual relations of a man and woman." On the other hand, Moshe Tendler, a professor of medical ethics at Yeshiva University, has concluded that there is religious authority for cloning, pointing out that respect for "sanctity of life would encourage us to use cloning if only for one individual...to prevent the loss of genetic line."

This is what we have come to expect from religious authorities: dogmatic pronouncements without any support external to a particular religious tradition, self-justifying appeals to a sect's teachings, and metaphor masquerading as reasoned argument. And, of course, the interpreters of God's will invariably fail to agree among themselves as to precisely what actions God would approve.

Given that these authorities have so little to offer by way of impartial, rational counsel, it would seem remarkable if anyone paid any attention to them. [See Richard Dawkins's "Thinking Clearly About Clones."] However, not only do these authorities have an audience, but their advice is sought out by the media and government representatives. Indeed, President Clinton's National Bioethics Advisory Commission devoted an entire day to hearing testimony from various theologians.

QUESTIONABLE ETHICS

The theologians' honored position reflects our culture's continuing conviction that there is a necessary connection between religion and morality. Most Americans receive instruction in morality, if at all, in the context of religious belief. As a result, they cannot imagine morality apart from religion, and when confronted by doubts about the morality of new developments in the sciences such as cloning—they invariably turn to their sacred writings or to their religious leaders for guidance. Dr. Ebbie Smith, a professor at Southwestern Baptist Theological Seminary, spoke to many Americans when he insisted that the Bible was relevant to the cloning debate because "the Bible contains God's revelation about what we ought to be and do, if we can understand it."

But the attempt to extrapolate a coherent, rationally justifiable morality from religious dogma is a deeply misguided project. [See Theodore Schick's "Morality Requires God...or Does It?"] To begin, as a matter of logic, we must first determine what is moral before we decide what "God" is telling us. As Plato pointed out, we cannot deduce ethics from "divine" revelation until we first determine which of the many competing revelations are authentic. To do that, we must establish which revelations make moral sense. Morality is logically prior to religion.

Moreover, most religious traditions were developed millennia ago, in far different social and cultural circumstances. While some religious precepts retain their validity because they reflect perennial problems of the human condition (for example, no human community can maintain itself unless basic rules against murder and stealing are followed), others lack contemporary relevance. The world of the biblical patriarchs is not our world. Rules prohibiting the consumption of certain foods or prescribing limited, subordinate roles for women might have some justification in societies lacking proper hygiene or requiring physical strength for survival. But they no longer have any utility and persist only as irrational taboos. In addition, given the limits of the world of the Bible and the Koran, their authors simply had no occasion to address some of the problems that confront us, such as the ethics of in vitro fertilization, genetic engineering, or cloning. To pretend otherwise, and to try to apply religious precepts by extension and analogy to these novel problems is an act of pernicious self-delusion.

To underscore these points, let us consider some of the more common objections to cloning that have been voiced by various religious leaders:

Cloning is playing god. This is the most common religious objection, and its appearance in the cloning debate was preceded by its appearance in the debate over birth control, the debate over organ transplants, the debate over assisted dying, etc. Any attempt by human beings to control and shape their lives in ways not countenanced by some religious tradition will encounter the objection that we are "playing God." To say that the objection is uninformative is to be charitable. The objection tells us nothing and obscures much. It cannot distinguish between interferences with biological process that are commonly regarded as permissible (for example, use of analgesics or antibiotics) and those that remain controversial. Why is cloning an impermissible usurpation of God's authority, but not the use of tetracycline?

Cloning is unnatural because it separates reproduction from human sexual activity. This is the flip side of the familiar religious objection to birth control. Birth control is immoral because it severs sex from reproduction. Cloning is immoral because it severs reproduction from sex. One would think that allowing reproduction to occur without all that nasty, sweaty carnal activity might appeal to some religious authorities, but apparently not. In any event, the "natural" argument is no less question-begging in the context of reproduction without sex than it is in the context of sex without reproduction. "Natural" most often functions as an approbative and indefinable adjective; it is a superficially impressive way of saying, "This is good, I approve." Without some argument as to why something is "natural" and "good" or "unnatural" or "bad," all we have is noise.

Cloning robs persons of their God-given uniqueness and dignity. Why? Persons are more than the product of their genes. Persons also reflect their experiences and relationships. Furthermore, this argument actually demeans human beings. It implies that we are like paintings or prints: the more copies that are produced, the less each is worth. To the contrary, each clone will presumably be valued as much by their friends, lovers, and spouses as individuals who are produced and born in the traditional manner and not genetically duplicated.

BEYOND THEOLOGY

All the foregoing objections assume that cloning could successfully be applied to human beings. It is worth noting that this issue is not entirely free from doubt since Dolly was produced only after hundreds of attempts. And although in principle the same techniques should work in humans, biological experiments cannot always be repeated across different species.

Of course, if some of the religious have their way, the general public may never know whether cloning would work in humans, as research into applications of cloning to human beings could be outlawed or driven underground. This would be an unfortunate development. Quite apart from the obvious, arguably beneficial, uses of cloning, such as asexual reproduction for those incapable of having children through sex, there are potential spinoffs from cloning research that could prove extremely valuable. Doctors, for example, could develop techniques to take skin cells from someone with liver disease, reconfigure them to function as liver cells, clone them, and then transplant them back into the patient. Such a procedure would avoid the sometimes-fatal complications that accompany genetically nonidentical transplants as well as problems caused by the chronic shortage of available organs for transplant.

This is not to discount the potential for harm and abuse that would result from the development of cloning technology, especially if we also master techniques for manipulating DNA. If we are able to modify a human being's genetic composition to achieve a predetermined end and can then create clones from the modified genetic structure, we could, theoretically, create a humanlike order of animals that would be more intelligent than other animals but less intelligent and more docile than (other?) human beings. Sort of ready-made slaves.

But religious precepts are neither necessary nor sufficient for avoiding such dangers. What we require is a secular morality based on our needs and interests and the needs and interests of other sentient beings. In considering the example just given, it is apparent that harmful consequences to normal human beings could result from the creation of these humanoid slaves, as many could be deprived of a means of earning their livelihood. It would also lead to an enormous and dangerous concentration of power in the hands of those who controlled these humanoids. And, although in the abstract we cannot decide what rights these humanoids would have, it is probable that, as sentient beings with at least rudimentary intelligence, they would have a right to be protected from ruthless exploitation and, therefore, we could not morally permit them to be treated as slaves. Even domesticated animals have a right to be protected from cruel and capricious treatment.

Obviously, I have not listed all the factors that would have to be considered in evaluating the moral implications of my thought experiment. I have not even tried to list all the factors that would have to be considered in assessing the many other ways—some of them now unimaginable—in which cloning technology might be applied. My point here is that we have a capacity to address these moral problems as they arise in a rational and deliberate manner if we rely on secular ethical principles. The call by many of the religious for an absolute ban on cloning experiments is a tacit admission that their theological principles are not sufficiently powerful and adaptable to guide us through this challenging future.

I want to make clear that I am not saying we should turn a deaf ear to those who offer us moral advice on cloning merely because they are religious. Many bioethicists who happen to have deep religious convictions have made significant, valuable contributions to this field of moral inquiry. They have done so, however, by offering secular and objective grounds for their arguments. Just as an ethicist's religious background does not entitle her to a special deference, so too her religious background does not warrant her exclusion from the debate, provided she appeals to reason and not supernatural revelation.

POLICY AND REGULATION

oliticians and governments have jumped right into the debate about cloning. We have seen that President Bill Clinton did not even wait for his own commission before he acted to ban federal support for work on human cloning. In this final section, we look at some of the responses to the possibility of human cloning by governments and their advisors. We follow with scholarly thoughts on societal attitudes and governmental involvement and responsibility.

President Clinton fit the norm for official reactions and actions. Governments and parliaments, and their various advisory bodies, were quick to express opposition to human cloning. Several years before Dolly, the European Parliament had stated its opposition to human cloning of any kind, calling on researchers "to pledge voluntarily not to engage in the cloning of human embryos for any purpose whatsoever." After Dolly, the stand was even stronger, stating bluntly that such cloning "cannot under any circumstances be justified or tolerated by any society." The advisors to the European Commission were likewise negatively inclined. Although they could see the justification of some animal cloning, in the case of humans any positive values from cloning (and they do seem to allow the possibility of these) would be outweighed by the bad or negative consequences. At least, they would in the case of reproductive human cloning where a new person was created. The advisors tread warily and seem not to want completely to bar research involving human cloning which is done "to throw light on the cause of human disease or to contribute to the alleviation of suffering."

At least part of the reason for the Parliament's absolutism—and remember that this body would be made up of elected representatives, as opposed to the commission's appointed experts—lies in Europe's horrific past, particularly in the perverted Nazi biological doctrines and practices. There is a terror of anything which even hints of a return to the 1930s, and you will note the Parliament's passing reference to the offensive nature of policies which permit "a eugenic and racist selection of the human race." The perversion of biology was not, of course, something on which Europeans have had an exclusive lien—eugenics was very popular in North America and racism is hardly absent today—and clearly past misuses were factors which influenced President Clinton's National Bioethics Advisory Commission. Its members also were against human cloning, calling for a continuation of Clinton's ban on the use of federal funding for work in this direction. Attempting such cloning would "at this time be an irresponsible, unethical, and unprofessional act."

Note however that the commission did not call for absolute bans on the use of cloning for research and the like, and they did not (much unlike the European Parliament) demand a total ban on all human reproductive cloning for all time. To the contrary, they stated strongly that any regulatory laws should have "sunset clauses," meaning that after a period of time (the commission mentions three to five years) the laws should wither away thus calling for new legislation, if at all. On what grounds would one then want to change things, perhaps allowing reproductive cloning? Obviously significant would be advances in the technology of animal cloning, so that dangers possibly attendant on human cloning would be reduced. But also, there is the question of public opinion. Clearly pertinent here are the different ethical and religious views that people have, and people's varied levels of sophistication about scientific knowledge. The commission felt that changes in official thinking must be responsive to changes in public thinking.

Concluding this section we have first Mary Warnock, a professional philosopher who has been much involved in British commissions on matters of bioethics. She notes explicitly how the Hitler-repeating-himself factor is at work in European reactions to cloning. Her message is that the sorts of moral issues raised by cloning cannot be simply solved on the basis of emotion, and that they call for calm, reflective judgment. One must get a proper balance between unfettered research—which is exciting and which may lead to great social benefits—and proper moral sensibilities. Regulation may well be necessary, but it should not be based on knee-jerk popular feeling. We must tread the line between "academic freedom" and "a democratic regard for the moral views of people at large."

American Daniel Callahan likewise takes up some of these issues, balancing the personal with the societal. He too sees that public opinion is a crucial factor in making governmental decisions about such things as cloning. There is a subtle interplay between advances in science and technology together with the thinking of professional ethicists on these issues, and in public opinion, which is a crucial factor in whether societies and their leaders will feel the need and the backing to intervene to regulate and legislate. Specifically in the case of cloning, Callahan suggests that today's initial negativism of parliaments and others to human cloning may well not prove to be definitive and lasting. Already, he sees a major change in general attitudes toward technology. There is now a much more positive public response, and people like Leon Kass (whom we met in Section IV) no longer speak for a majority or general position. Also, Callahan sees at large a much stronger public feeling that people have the absolute right to reproduction, and if this includes cloning then so be it. He sees that today's society much more clearly and definitively defines as one of its goals the need to overcome infertility. This last point, Callahan links to the fact that in today's (American) society infertility is a more pressing problem than hitherto, thanks to such things as the tendency toward later female reproduction. This is a phenomenon brought about in part by major social changes in the status of women.

Callahan himself admits that he himself is not necessarily keen on these societal changes toward reproductive technology, but recognizes that they must be faced. And on this point we have a good place to end our readings. We have seen again and again how cloning, plants and animals in part and humans in whole, raises serious social and ethical questions. The scientists and technicians who do the cloning, the ethicists and theologians and religious leaders who think and talk about the cloning, all live in societies which must either permit or ban or regulate the cloning. There are tensions and difficult-to-solve questions about how to bring into harmony the interests and actions and thoughts of the few with the beliefs and needs and fears and emotions of the many. We now invite you to think about these matters and to start to work toward your own conclusions.

Cloning Animals and Human Beings

The European Parliament

RESOLUTION ON CLONING

The European Parliament,

—having regard to the alarm caused by the announcement on 24 February 1997 from the Roslin Institute and Pharmaceutical Proteins Ltd of Scotland of the production of a sheep cloned from an adult cell and the possibility of such reproductive techniques being used to produce human embryos,

—having regard to its resolutions of 16 March 1989 on the ethical and legal problems of genetic engineering[1] and artificial insemination "in vivo" and "in vitro"[2] and 28 October 1993 on the cloning of human embryos,[3]

—having regard to the Council of Europe's Convention for the Protection of Human Rights and Dignity of the Human Being with regard to the Application of Biology and Medicine ("Bioethics Convention"),[4] and Parliament's resolution of 20 September 1996 thereon,[5]

From *Bulletin of Medical Ethics* 128 (May 1997): 10–11. Reprinted with permission.

The European Parliament: Cloning Animals and Human Beings

—having regard to the reports of the Commission's ethical advisory group on biotechnology,

—having regard to recommendation 1046 of the Parliamentary Assembly of the Council of Europe on the use of human embryos,[6]

—confirming its opposition to the cloning of human embryos, the position it adopted in its resolutions of 1989 and 1993 referred to above,

A. whereas cloning breaks new ethical ground and has led to great public concern,

B. in the clear conviction that the cloning of human beings, whether experimentally, in the context of fertility treatment, preimplantation diagnosis, tissue transplantation, or for any other purpose whatsoever, cannot under any circumstances be justified or tolerated by any society, because it is a serious violation of fundamental human rights and is contrary to the principle of equality of human beings as it permits a eugenic and racist selection of the human race, it offends against human dignity and it requires experimentation on humans,

C. whereas there is a need to ensure that the benefits of biotechnology are not lost as a result of sensationalist and alarmist information,

D. whereas adequate methods of regulating and policing developments in the field of genetics must be established,

E. whereas all necessary information must be made available to the public, and the EU must now take the lead in promoting full public consideration of these questions,

F. whereas the Convention on Human Rights and Biomedicine does not expressly ban the cloning of human beings, and in any event is not yet in force in any EU Member State,

G. whereas some Member States have no national legislation prohibiting the cloning of human beings,

H. whereas cloning of humans for all purposes should be banned in the EU,

I. whereas international action is required,

1. Stresses that each individual has a right to his or her own genetic identity and that human cloning is, and must continue to be, prohibited;

2. Calls for an explicit worldwide ban on the cloning of human beings;

3. Urges the Member States to ban the cloning of human beings at all stages of formation and development, regardless of the method used, and to provide for penal sanctions to deal with any violation;

4. Calls on the Commission to report to it on any research carried out in this field on Community territory and on the legal framework existing in the Member States;

5. Calls on the Commission to check whether human cloning could form part of research programs financed by the Community and, if so, to block the appropriations for them;

6. Believes it is essential to establish ethical standards, based on respect for human dignity, in the areas of biology, biotechnology, and medicine;

7. Believes that it is desirable for such standards to apply globally and that they should conform to a high level of protection;

8. Considers that the direct protection of the dignity and rights of individuals is of absolute priority as compared with any social or third-party interest;

9. Calls for the establishment of a European Union Ethics Committee to assess ethical aspects of applications of gene technology and to monitor developments in this field; calls on the Commission to submit proposals for the composition and terms of reference of the committee under the procedure set out in Article 189b of the EC Treaty, while ensuring that it is constituted with full respect for transparency and democratic principles, and that all appropriate interested groups are represented;

10. Considers that, in view of the universality of the principles relating to the dignity of the human being, efforts must be made by the European Union, its Member States, and the United Nations to promote worldwide governance on this issue and to promote and put into effect binding international agreements in order to ensure that such principles are applied worldwide;

11. Calls on researchers and doctors engaged in research on the human genome to abstain spontaneously from participating in the cloning of human beings until the entry into force of a legally binding ban;

12. Acknowledges that research in the field of biotechnology, in particular the manufacture of proteins, medicines, and vaccines for human use, could help to combat certain diseases;

13. Considers that the international scientific community and governments should provide the public with complete and up-to-date information on current research relating to gene technology;

14. Calls on the Commission, in connection with its research programs, to prepare a recommendation concerning bioethics laying down strict limits on its research in accordance with respect for human life and to consider, if necessary, whether action at Community level is necessary;

15. Calls on the Commission to propose Community legislation on animal cloning and in particular on the new scientific developments, with strict controls to guarantee human health and the continuation of animal species and races and to safeguard biological diversity;

16. Instructs its President to forward this resolution to the Commission, the Council, the governments of the Member States, the Secretary-General and Parliamentary Assembly of the Council of Europe and the Secretary-General of the United Nations.

NOTES

1. OJ C 96, 17.4.1989, p. 165.
2. OJ C 96, 17.4.1989, p. 171.
3. OJ C 315, 22.11.1993, p. 224.
4. Adopted by the Committee of Ministers on 19 November 1996, document DIR/JUR(96)14 of the Directorate of Legal Affairs of the Council of Europe.
5. OJ C 320, 28.11.1996, p. 268.
6. Recommendation 1046(1986) adopted on 24 September 1986 (18th sitting).

Ethical Aspects of Cloning Techniques

(26)

Advisors to the President of the European Commission on the Ethical Implications of Biotechnology

EDITOR'S NOTE

Following a request from the President of the European Commission, M Jacques Santer, the Group of Advisers on the Ethical Implications of Biotechnology adopted an Opinion on 28 May 1997, examining the ethical consequences of cloning. The Opinion is published in its entirety here.

The Group of Advisers on the Ethical Implications of Biotechnology (GAEIB) to the European Commission,

Having regard to the commission's request of 28 February 1997 for an Opinion on the ethical implications of cloning techniques, namely animal cloning and application potential to human beings,

Reprinted with permission of BMJ Publishing Group from *Journal of Medical Ethics* 23 (1997): 349–52.

Having regard to the Treaty on European Union, namely the article F 2 of the Common Provisions and the annexed Declaration no. 24 on the Protection of Animals,

Having regard to the council directive 86/609/EEC regarding the protection of animals used for experimental and other scientific purposes,

Having regard to the council directive 90/220/EEC, regarding the deliberate release into the environment of genetically modified organisms,

Having regard to the council and European parliament decision no. 1110/94/EC of 26 April 1994 adopting the 4th Framework Programme,

Having regard to the resolutions of the European parliament, namely the resolutions of 16 March 1989 on the ethical and legal problems of genetic engineering, of 28 October 1993 on the cloning of the human embryo and of 12 March 1997 on cloning,

Having regard to the European Conventions of the Council of Europe for the protection of animals kept for farming purposes (1976-EST 87) and in particular the protocol of amendment thereto, and for the protection of vertebrate animals used for experimental or other scientific purposes (1986-EST 123),

Having regard to the Convention of the Council of Europe on Human Rights and Biomedicine, signed on 4th April 1997,

Having regard to the United Nations Convention on Biodiversity of 6 June 1992, ratified by the European Union on 25 October 1993,

Having regard to the draft declaration of UNESCO, Universal Declaration on the Human Genome and Human Rights, of 20 December 1996,

Having regard to the hearings, organized on 18 April 1997 by the GAEIB with members of the European parliament and commission, international organizations (WHO and UNESCO), researchers, industry, representatives of consumers, patients and environment organizations and animal protection associations.

The following points aim to shed light on the cloning debate by giving information on the scientific aspects of cloning and the ethical problems relating to them.

1. WHEREAS

1.1 Cloning is the process of producing "genetically identical" organisms. It may involve division of a single embryo, in which case both the nuclear genes and the small number of mitochondrial genes would be "identical," or it may involve nuclear transfer, in which case only the nuclear genes would be "identical." But

CLONING

genes may be mutated or lost during the development of the individual: the gene set may be identical but it is unlikely that the genes themselves would ever be totally identical. In the present context, we use the term "genetically identical" to mean "sharing the same nuclear gene set."

1.2 It is inherent in the process of sexual reproduction that the progeny differ genetically from one another. In contrast, asexual reproduction (cloning) produces genetically identical progeny. This is a common form of reproduction in plants, both in nature and in the hands of plant breeders and horticulturalists. Once a desired combination of characteristics has been achieved asexual reproduction is the best way of preserving it. Asexual reproduction is also common among some invertebrate animals (worms, insects). Asexual reproduction in plants and inverte-brates usually takes place by budding or splitting.

1.3 The first successful cloning in vertebrate animals was reported in 1952, in frogs. Nuclei from early frog embryos were transferred to unfertilized frog eggs from which the original nuclei had been removed. The resulting clones were not reared beyond the tadpole stage. In the 1960s, clones of adult frogs were produced by transfer not only of nuclei from early embryos but also of nuclei from differ-entiated larval intestinal cells. Later, clones of feeding tadpoles were obtained by nuclear transfer from differentiated adult cells, establishing that differentiation of cells involving selective gene expression does not require the loss or irreversible inactivation of genes. Nuclear transfer in frogs has not yet generated an adult animal from cells of an adult animal.

1.4 Nuclear transfer can be used for different objectives. Nuclear transfer in mice has been used to show that both a female and male set of genes are required for development to birth. If the two pronuclei, taken from fertilized eggs and trans-ferred into an enucleated egg, are only maternal or only paternal, normal devel-opment does not occur. This is not cloning, since the single embryo formed is not identical to any other embryo and the objective is not to multiply individuals.

1.5 Nuclear transfer has also been used for cloning in various mammalian species (mice, rabbits, sheep, cattle), but until recently only nuclei taken from very early embryos were effective, and development was often abnormal, for reasons that are not fully understood.

1.6 In contrast, cloning by embryo splitting, from the 2-cell up to the blastocyst stage, has been extensively used in sheep and cattle to increase the yield of progeny from genetically high-grade parents. Because of the different pattern of early development, embryo splitting is much less successful in mice. From a scientific point of view, it would probably not be very effective in the human, although monozygotic (one-egg) twins and higher multiples occur naturally at a low incidence.

1.7 In 1996, a new method of cloning sheep embryos was reported, which involved first establishing cell cultures from single embryos. Nuclei from the cultured cells were transferred to enucleated unfertilized sheep eggs, particular attention being paid to the cell cycle stage of both donor and host cells, and the eggs were then artificially stimulated to develop. Genetically identical normal lambs were born.

1.8 Cell cultures were then established not only from embryonic and fetal stages, but also from mammary tissue taken from a six-year-old sheep. Nuclear transfer was carried out as before, and in 1997 it was reported that several lambs had been born from the embryonic and fetal transfers and one lamb named Dolly (out of 277 attempts) from the adult nuclear transfer. It is not known whether the transferred nucleus was from a differentiated mammary gland cell or from a stem cell.

1.9 From the point of view of basic research, this result is important. If repeatable it may allow greater insight into the aging process, how much is due to cell aging, and whether or not it is reversible. Such work may also increase our understanding of cell commitment, the origin of the cancer process, and whether it can be reversed, but at the present time the research is at a very early stage. Dolly may have a shortened life span or a greater susceptibility to cancer: if she is fertile, her progeny may show an increased abnormality rate, owing to the accumulation of somatic mutations and chromosomal damage.

Concerning the Applications of Animal Cloning

1.10 Potential uses of cloning animals are reported to include:

—in the field of medicine and medical research, to improve genetic and physiological knowledge, to make models for human diseases, to produce at lower cost proteins like milk proteins to be used for therapeutic aims, to provide a source of organs or tissues for xenotransplantation;

—in agriculture and agronomical research, to improve the selection of animals or to reproduce animals having specific qualities (longevity, resistance,...) either innate, or acquired by transgenesis.

CLONING

1.11 From the point of view of animal breeding, the technology could be useful, in particular if it increases the medical and agricultural benefits expected from transgenesis (genetic modification of animals). By using genetic modification and selection in cultured cell lines, rather than in adult animals, it could become possible to remove genes, such as those provoking allergic reactions, as well as adding genes, for the benefit of human health.

1.12 Furthermore, transgenesis is an uncertain process: different transgenic animals express the introduced gene in a different manner and to a different extent, and do not "breed true." Cloning of adult animals of high performance, if it is possible, would reduce the number of transgenic animals needed and would allow human pharmaceuticals, for example, to be produced at a lower cost than would be possible otherwise.

1.13 If the use of cloning became more widespread in the animal breeding industry, for example to bring the level of the general herd up to the level of the elite breeding populations, there is a danger that the level of genetic diversity could fall to an unacceptable degree. The introduction of artificial insemination in cattle raised similar problems.

Concerning Human Implications

1.14 A clear distinction must be drawn between reproductive cloning aimed at the birth of identical individuals, which in humans has never been performed, and non-reproductive cloning, limited to the in vitro phase.

1.15 In considering human implications, we must again distinguish between cloning by embryo splitting and cloning by nuclear replacement (see 1.1). We must also distinguish between nuclear replacement as a means of cloning and nuclear replacement as a therapeutic measure, for example to avoid the very serious consequences of mitochondrial disease. The latter situation, which would require an enucleated donor egg containing normal mitochondria, as it need not involve the production of genetically identical individuals, will not be considered further here although we appreciate that it will raise ethical problems of its own.

1.16 Embryo splitting in the human is the event that gives rise to monozygotic (one-egg) twins and higher multiples. It has been discussed in the context of assisted reproduction, as a means of increasing the success rate of IVF, but there is no evidence that it has ever been used for this purpose, nor that it would be effective if it were so used, because of the pattern of early development of the human embryo.

1.17 Monozygotic twins show us that genetically identical individuals are far from identical: they may differ from one another not only physically but also psychologically, and in terms of personality. Individuals cloned by nuclear transfer from an adult cell would of course be even more different from their donor, since they would have different mitochondrial populations, they would be different in age and they would have had a different environment both before and after birth and a different upbringing. We are not just our genes.

1.18 There is no ethical objection to genetically identical human beings per se existing, since monozygotic twins are not discriminated against. However, the use of embryo splitting, or the use of human embryo cells as nuclear donors, deliberately to produce genetically identical human beings raises serious ethical issues, concerned with human responsibility and instrumentalization of human beings.

1.19 However, research involving human nuclear transfer could have important therapeutic implications, for example the development of appropriate stem cell cultures for repairing human organs. It could also provide insights into how to induce regeneration of damaged human tissues. If such research resulted in embryonic development, serious and controversial ethical issues concerning human embryo research would of course arise. Any attempt to develop methods of human reproductive cloning would require a large amount of human experimentation.

1.20 If adult cells were to be used as nuclear donors, we are still ignorant of the possible risks: whether the cloned individuals would have a shorter life span, a greater susceptibility to cancer, whether they would be fertile, and if so whether they or their offspring would suffer from an abnormal rate of genetic abnormalities. Furthermore, the procedure would be immensely costly: each attempt would require several eggs and an available uterus, and many attempts would be unsuccessful. The issues of human responsibility and instrumentalization of human beings are even more ethically acute in this context.

CLONING

2. The Group Submits the Following Opinion to the European Commission

Concerning Cloning of Animals

2.1 Research on cloning in laboratory and farm animals is likely to add to our understanding of biological processes, in particular aging and cell commitment, and hence may contribute to human well-being. It is ethically only acceptable if carried out with strict regard to animal welfare, under the supervision of licensing bodies.

2.2 Cloning of farm animals may prove to be of medical and agricultural as well as economic benefit. It is acceptable only when the aims and methods are ethically justified and when it is carried out under ethical conditions, as outlined in the GAEIB's Opinion no. 7 on the Genetic Modification of Animals.

2.3 These ethical conditions include:

—the duty to avoid or minimize animal suffering since unjustified or disproportionate suffering is unacceptable;

—the duty of reducing, replacing, and when possible refining the experimentation adopted for the use of animals in research;

—the lack of better alternatives;

—human responsibility for animals, nature, and the environment, including biodiversity.

2.4 Particular attention should be paid to the need to preserve genetic diversity in farm animal stocks. Strategies to incorporate cloning into breeding schemes while maintaining diversity should be developed by European institutions.

2.5 Insofar as cloning contributes to health, special attention should be paid to the public's right to protection against risks as well as their right to adequate information. Furthermore, if the costs of production are reduced, consumers should also benefit.

Concerning Human Implications

2.6 As far as reproductive cloning is concerned, many motives have been proposed, from the frankly selfish (the elderly millionaire vainly seeking immortality) to the apparently acceptable (the couple seeking a replacement for a dead child, or a fully compatible donor for a dying child, or the attempt to perpetuate some extraordinary artistic or intellectual talent). Considerations of instrumentalization and eugenics render any such acts ethically unacceptable. In addition, since these techniques entail increased potential risks, safety considerations constitute another ethical objection. In the light of these considerations, any attempt to produce a genetically identical human individual by nuclear substitution from a human adult or child cell ("reproductive cloning") should be prohibited.

2.7 The ethical objections against cloning also rule out any attempt to make genetically identical embryos for clinical use in assisted reproduction, either by embryo splitting or by nuclear transfer from an existing embryo, however understandable.

2.8 Multiple cloning is a fortiori unacceptable. In any case, its demands on egg donors and surrogate mothers would be outwith the realms of practicality at the present time.

2.9 Taking into account the serious ethical controversies surrounding human embryo research: for those countries in which nontherapeutic research on human embryos is allowed under strict license, a research project involving nuclear substitution should have the objective either to throw light on the cause of human disease or to contribute to the alleviation of suffering, and should not include replacement of the manipulated embryo in a uterus.

2.10 The European Community should clearly express its condemnation of human reproductive cloning and should take this into account in the relevant texts and regulations in preparation as the Decision adopting the Vth Framework Programme for Research and Development (1998–2002) and the proposed directive on legal protection of biotechnological inventions.

General Remarks

2.11 Further efforts must be made to inform the public, to improve public awareness of potential risks and benefits of such technologies, and to foster informed opinion. The European Commission is invited to stimulate the debate involving the public, consumers, patients, environment and animal protection associations, and a well-structured public debate should be set up at European level. Universities and high schools should also be involved in the debate at European level.

2.12 These new technologies increase the power of people over nature and thus increase their responsibilities and duties. Along the line of the promotion by the European Commission of research on the ethical, legal, and social aspects of life sciences, the commission should continue to foster ethical research on cloning related areas, at a European level.

In accordance with its mandate, the Group of Advisers on the Ethical Implications of Biotechnology submits this Opinion to the European Commission.

The members: Anne McLaren, Margareta Mikkelsen, Luis Archer, Octavi Quintana, Stefano Rodota, Egbert Schroten, Dietmar Mieth, Gilbert Hottois; Chairman, Noëlle Lenoir.

Cloning Human Beings
Executive Summary
National Bioethics Advisory Commission

(27)

I n February of this year two figures were added to our daily life. Dolly, a cloned sheep, and her maker, Scottish scientist Ian Wilmut, could be found in our newspapers, on our televisions, across the Internet, and in our conversations. What Wilmut had done was indeed new, if not fantastic. By transferring the nucleus of a somatic cell from an adult animal into an egg from which the nucleus had been removed, Wilmut successfully cloned a mammal—a technique that had never before succeeded.

The drama of this scientific capability, and its potential reach toward human beings, prompted President Clinton to ask his newly formed National Bioethics Advisory Commission to spend ninety days examining the ethical and legal issues raised by the possibility of cloning human beings. This collection of essays responds to NBAC's report.

Though the science that made Dolly possible is unprecedented, two authors remind us that much of this debate is not new. By the late 1960s and 1970s, scholars Leon Kass, Paul Ramsey, and others had already started to speculate, and

Reprinted from *Hastings Center Report* (September–October 1997): 6–9.

worry, about the possibility of cloning human beings. Though they did not have the details of Wilmut's technique, which NBAC called "somatic cell nuclear transfer" (SCNT), to guide their analyses, they nonetheless foresaw with clarity many of the issues.

And, according to some, the ethical issues are profound. Reiterating the report, one author clearly lays out those raised by SCNT. Parents hoping to clone using Wilmut's technique can "circumvent the chance events of meiosis and fertilization" and enter the baby-making enterprise with a tremendous amount of control over and knowledge about their future child's characteristics. Both the prospect for increased control and knowledge are worrisome, particularly for those who believe, as one author says, that the practice of cloning children threatens the "moral significance of the family, the meaning of parenthood, and the ethics of unchosen obligations."

Both prospects are possible because the technique uses *adult* cells. Indeed, NBAC's analysis hangs on the subtle but crucial distinction between adult and embryonic or fetal cells. While one author lauds the commissioner's attempts to make clear the boundaries of their analysis and recommended temporary ban— only to techniques using adult cells aimed at producing a child—another thinks the commissioners have been anything but clear. The ban is overly broad and overly vague, likely to chill important research both inside and outside these parameters, and for an unlimited time. And the ban's "sunset" provision fails to satisfy. Some worry that Congress will never muster the political will to lift the ban after the specified three to five years; others, to keep it in place. Much will depend on the "national dialogue" that NBAC has urged us to undertake.—*EB*

Executive Summary

From Cloning Human Beings: The Report and Recommendations of the National Bioethics Advisory Commission
(Rockland, Md., June 1997)

The idea that humans might someday be cloned—created from a single somatic cell without sexual reproduction—moved further away from science fiction and closer to a genuine scientific possibility on February 23, 1997. On that date, the *Observer* broke the news that Ian Wilmut, a Scottish scientist, and his colleagues at the Roslin Institute were about to announce the successful cloning of a sheep by a technique which had never before been fully successful in mammals. The technique involved transplanting the generic material of an adult sheep, apparently obtained from a differentiated somatic cell, into an egg from which the nucleus had been removed. The resulting birth of the sheep, named Dolly, on July 5, 1996, was different from prior attempts to create identical offspring since Dolly contained the generic material of only one parent, and was, therefore, a "delayed" genetic twin of a single adult sheep.

This cloning technique is an extension of research that had been ongoing for over forty years using nuclei derived from nonhuman embryonic and fetal cells. The demonstration that nuclei from cells derived from an adult animal could be "reprogrammed," or that the full generic complement of such a cell could be reactivated well into the chronological life of the cell, is what sets the results of this experiment apart from prior work. In this report we refer to the technique, first described by Wilmut, of nuclear transplantation using nuclei derived from somatic cells other than those of any embryo or fetus as "somatic cell nuclear transfer."

Within days of the published report of Dolly, President Clinton instituted a ban on federal funding related to attempts to clone human beings in this manner. In addition, the President asked the recently appointed National Bioethics Advisory Commission (NBAC) to address within ninety days the ethical and legal issues that surround the subject of cloning human beings. This provided a welcome opportunity for initiating a thoughtful analysis of the many dimensions of the issue, including a careful consideration of the potential risks and benefits. It also presented an occasion to review the current legal status of cloning and the potential constitutional challenges that might be raised if new legislation were enacted to restrict the creation of a child through somatic cell nuclear transfer cloning.

The Commission began its discussion fully recognizing that any effort in humans to transfer a somatic cell nucleus into an enucleated egg involves the creation of an embryo, with the apparent potential to be implanted in utero and developed to term. Ethical concerns surrounding issues of embryo research have recently received extensive analysis and deliberation in our country. Indeed, federal funding for human embryo research is severely restricted, although there are few restrictions on human embryo research carried out in the private sector. Thus, under current law, the use of somatic cell nuclear transfer to create an embryo solely for research purposes is already restricted in cases involving federal funds. There are, however, no current federal regulations on the use of private funds for this purpose.

The unique prospect, vividly raised by Dolly, is the creation of a new individual genetically identical to an existing (or previously existing) person—a "delayed" genetic twin. This prospect has been the source of the overwhelming public concern about such cloning. While the creation of embryos for research purposes alone always raises serious ethical questions, the use of somatic cell nuclear transfer to create embryos raises no new issues in this respect. The unique and distinctive ethical issues raised by the use of somatic cell nuclear transfer to create children relate to, for example, serious safety concerns, individuality, family integrity, and treating children as objects. Consequently, the Commission focused its attention on the use of such techniques for the purpose of creating an embryo which would then be implanted in a woman's uterus and brought to term. It also expanded its analysis of this particular issue to encompass activities in both the public and private sector.

In its deliberations, NBAC reviewed the scientific developments which preceded the Roslin announcement, as well as those likely to follow in its path. It also considered the many moral concerns raised by the possibility that this technique could be used to clone human beings. Much of the initial reaction to this possibility was negative. Careful assessment of that response revealed fears about harms to the children who may be created in this manner, particularly psychological harms associated with a possibly diminished sense of individuality and personal autonomy. Others expressed concern about a degradation in the quality of parenting and family life.

In addition to concern about specific harms to children, people have frequently expressed fears that the widespread practice of somatic cell nuclear transfer cloning would undermine important social values by opening the door to a form of eugenics or by tempting some to manipulate others as if they were objects instead of persons. Arrayed against these concerns are other important social values, such as protecting the widest possible sphere of personal choice, particularly in matters pertaining to procreation and child rearing, maintaining privacy and the freedom of scientific inquiry, and encouraging the possible development of new biomedical breakthroughs.

To arrive at its recommendations concerning the use of somatic cell nuclear transfer techniques to create children, NBAC also examined long-standing religious traditions that guide many citizens' responses to new technologies and found that religious positions on human cloning are pluralistic in their premises, modes of argument, and conclusions. Some religious thinkers argue that the use of somatic cell nuclear transfer cloning to create a child would be intrinsically immoral and thus could never be morally justified. Other religious thinkers contend that human cloning to create a child could be morally justified under some circumstances, but hold that it should be strictly regulated in order to prevent abuses.

The public policies recommended with respect to the creation of a child using somatic cell nuclear transfer reflect the Commission's best judgments about both the ethics of attempting such an experiment and our view of traditions regarding limitations on individual actions in the name of the common good. At present, the use of this technique to create a child would be a premature experiment that would expose the fetus and the developing child to unacceptable risks. This in itself might be sufficient to justify a prohibition on cloning human beings at this time, even if such efforts were to be characterized as the exercise of a fundamental right to attempt to procreate.

Beyond the issue of the safety of the procedure, however, NBAC found that concerns relating to the potential psychological harms to children and effects on the moral, religious, and cultural values of society merited further reflection and deliberation. Whether upon such further deliberation our nation will conclude that the use of cloning techniques to create children should be allowed or permanently banned is, for the moment, an open question. Time is an ally in this regard, allowing for the accrual of further data from animal experimentation, enabling an assessment of the prospective safety and efficacy of the procedure in humans, as well as granting a period of fuller national debate on ethical and social concerns. The Commission therefore concluded that there should be imposed a period of time in which no attempt is made to create a child using somatic cell nuclear transfer.[1]

Within this overall framework the Commission came to the following conclusions and recommendations:

I. The Commission concludes that at this time it is morally unacceptable for anyone in the public or private sector, whether in a research or clinical setting, to attempt to create a child using somatic cell nuclear transfer cloning. We have reached a consensus on this point because current scientific information indicates that this technique is not safe to use in humans at this time.

Indeed, we believe it would violate important ethical obligations were clinicians or researchers to attempt to create a child using these particular technologies, which are likely to involve unacceptable risks to the fetus and/or potential child. Moreover, in addition to safety concerns, many other serious ethical concerns have been identified, which require much more widespread and careful public deliberation before this technology may be used.

The Commission, therefore, recommends the following for immediate action:

- A continuation of the current moratorium on the use of federal funding in support of any attempt to create a child by somatic cell nuclear transfer.
- An immediate request to all firms, clinicians, investigators, and professional societies in the private and nonfederally funded sectors to comply voluntarily with the intent of the federal moratorium. Professional and scientific societies should make clear that any attempt to create a child by somatic nuclear transfer and implantation into a woman's body would at this time be an irresponsible, unethical, and unprofessional act.

CLONING

II. The Commission further recommends that:

- Federal legislation should be enacted to prohibit anyone from attempting, whether in a research or clinical setting, to create a child through somatic cell nuclear cloning. It is critical, however, that such legislation include a sunset clause to ensure that Congress will review the issue after a specified time period (three to five years) in order to decide whether the prohibition continues to be needed. If state legislation is enacted, it should contain such a sunset provision. Any such legislation or associated regulation also ought to require that some point prior to the expiration of the sunset period, an appropriate oversight body will evaluate and report on the current status of somatic cell nuclear transfer technology and on the ethical and social issues that its potential use to create human beings would raise in light of public understandings at that time.

III. The Commission also concludes that:

- Any regulatory or legislative actions undertaken to effect the foregoing prohibition on creating a child by somatic cell nuclear transfer should be carefully written so as not to interfere with other important areas of scientific research. In particular, no new regulations are required regarding the cloning of human DNA sequences and cell lines, since neither activity raises the scientific and ethical issues that arise from the attempt to create children through somatic cell nuclear transfer, and these fields of research have already provided important scientific and biomedical advances. Likewise, research on cloning animals by somatic cell nuclear transfer does not raise the issues implicated in attempting to use this technique for human cloning, and its continuation should only be subject to existing regulations regarding the humane use of animals and review by institution-based animal protection committees.
- If a legislative ban is not enacted, or if a legislative ban is ever lifted, clinical use of somatic cell nuclear transfer techniques to create a child should be preceded by research trials that are governed by twin protections of independent review and informed consent, consistent with existing norms of human subjects protection.
- The United States Government should cooperate with other nations and international organizations to enforce any common aspects of their respective policies on the cloning of human beings.

IV. The Commission also concludes that different ethical and religious perspectives and traditions are divided on many of the important moral issues that surround any attempt to create a child using somatic cell nuclear transfer techniques. Therefore, we recommend that:

- The federal government, and all interested and concerned parties, encourage widespread and continuing deliberation on these issues in order to further our understanding of the ethical and social implications of this technology and to enable society to produce appropriate long-term policies regarding this technology should the time come when present concerns about safety have been addressed.

V. Finally, because scientific knowledge is essential for all citizens to participate in a full and informed fashion in the governance of our complex society, the Commission recommends that:

- Federal departments and agencies concerned with science should cooperate in seeking out and supporting opportunities to provide information and education to the public in the areas of genetics, and on other developments in the biomedical sciences especially where these affect important cultural practices, values, and beliefs.

Reference

1. The Commission also observes that the use of any other technique to create a child genetically identical to an existing (or previously existing) individual would raise many, if not all, of the same non-safety-related ethical concerns raised by the creation of a child by somatic cell nuclear transfer.

The Regulation of Technology
28
Mary Warnock

Everybody recognizes that most of the problems in medical ethics arise, these days, from innovations in medical technology. We would not have had to lay down laws or ethical guidelines about assisted reproduction had it not been for the new technology of in vitro fertilization, which produced the first IVF baby in 1978. We would not be currently anxious about the ethics of possible human cloning, had it not been for the production in Edinburgh of Dolly, the lamb whose birth resulted from the removal of a mammary gland cell from an adult sheep. So the question is whether there is some research into developing technology that is too dangerous, that will lead to consequences too dramatic for humanity, for the research itself to be permitted. Should there be control over what technological innovation should be permitted?

Put like this, the question looks absurd. It is not the discovery of new technological possibilities that is alarming, but the use to which these possibilities may

Reprinted from *Cambridge Quarterly of Healthcare Ethics* 7, no. 2 (Spring 1998): 173–75. Copyright © 1998 Cambridge University Press. Reprinted with the permission of Cambridge University Press.

be put. Control should not be over research, but over the uses of research. After all, even Plato, centuries ago, recognized that any skill, or *techne*, could be put to either good or bad use; the skilled doctor could also be a skilled poisoner.

However, the distinction between research and the uses of research is by no means easy to draw. First, it may be argued that if a procedure is shown to be possible (such as, for example, the transplant of organs from one human to another, or, transgenically, from one animal to another), then someone, somewhere, will want to use this technique for therapeutic, not merely for research purposes. Second, the very possibility of such a technique may have been established only by means of its use on subjects, animal or human. There is no way of definitively distinguishing new and untried treatment from research. All treatment is, in some sense, a contribution to research, or may be such. Equally, in the field of medicine all research is undertaken with at least a vague hope that it will one day be used to improve treatment. Medical research is seldom entirely "pure." So the development of new technology cannot be fenced off from its use.

Nevertheless, it could be argued that the development of certain techniques is simply in itself too dangerous to be permitted. In the 1970s, when the genetic manipulation of plants became a widely recognized possibility, a moratorium was, for a time, called on such research, on the grounds that the research itself was too dangerous, carrying as it did a risk to those engaged in it, and a risk of the accidental release into the environment of genetically modified organisms, with unknown consequences. The moratorium did not last; and it is probably true to say that the safety of research workers and of the environment as a whole is better protected than it was because of a greater realization of the risks that may exist unless due care is taken. So the dangers of research are these days seen to be dangers of outcome rather than of the processes themselves. A parallel story could be told of the fears surrounding research in nuclear physics.

Thus the question must be asked again, are there some technologies to develop which would be so threatening that they should be subject to regulation, or even be prohibited by law? The technique of cloning is obviously a candidate for such prohibition. The public reaction to the birth of the lamb, Dolly, was little short of hysterical. On Sunday, 23 February 1997, the *Observer* carried the story of the cloning, to be published with proper scientific dispassion in *Nature* the following Thursday. The press reacted instantly, both in the United Kingdom and abroad. For some reason, Philippe Vasseur, the French minister of Agriculture, warned Europe of the possibility of six-legged chickens. But, unsurprisingly, most concentrated on the possible use of the technique on humans. The German newspaper *Die Welt* called attention to the political implications of human cloning, saying that Hitler would have used it if it had then been possible; Jacques Santer, the president of the European Commission, instructed Commission officials to investigate whether there was need for EU regulation of cloning; and the German Socialist MEP, Dagmar Roth-Behrend, called for a worldwide moratorium on the technique, on whatever animals it was used. Fortunately, no one rushed out instant regulative legislation. The scientific press managed to come up with explanatory and generally reassuring accounts of the procedure, and the Edinburgh team themselves very sensibly announced that they were not in favor of the use of the technique on humans.

There is, however, a lesson to be learned from the case of cloning. It is tempting for the press (and they are certain to fall for the temptation) to turn the announcement of any new biomedical technique into a shock/horror story; and the public will probably accept what they read and put pressure on Parliament to take steps either to prohibit further research altogether or at least to subject it to nonscientific regulation. To legislate in such circumstances, in response to popular feeling, is almost always a mistake. But in any case there is a fundamental objection to the regulation of scientific research, and in the excitement of the moment it must not be forgotten. It is the need to preserve academic freedom. By this I do not mean an absolute right of scientists or other academics to receive public funding for whatever they want to do, or to teach, I mean rather that academics themselves must be recognized as those who can decide what is or is not worth pursuing. Research, or indeed the content of teaching for that matter, must not be controlled by those who are ignorant. Parliament, and the general public, must trust those who actually know what they are talking about, and must be taught by them. We are all too likely to think that anybody is entitled to hold a moral view, either about what research is or is not worth pursuing, or about what the possible outcomes of such research may be. But this is a false belief; it is not possible to hold a responsible moral opinion on a matter of which one is ignorant. We need to learn the facts, and the probabilities, first, and then form a judgment upon them. Legislation based on popular indignation or fear, then, is nearly always going to be bad legislation that will be later regretted.

However, this is not to say that technology and the search for new technologies must never be subject to legislation. If only to allay public alarm (fear, that is, that scientists are too powerful, and that they like to "play God") it is often necessary that the use of technology should be, if not prohibited in certain cases, then at least regulated. And if necessary the criminal law must be invoked in the case of nonobservance of regulation. For example, so horrendous did people find the idea of fertilizing sperm and egg in the laboratory, and keeping the resulting embryo alive in its "test-tube" indefinitely, that a new criminal offense was invented in the legislation of 1990—that of keeping an embryo alive for more than fourteen days after the completion of fertilization, an offense that carries the penalty of up to ten years imprisonment. Some would argue that the creating of human clones by the technique that produced Dolly should likewise become a criminal offense, though I believe that this is unnecessary, at least for the foreseeable future.

If regulation of the uses of new technology is ever to be thought desirable, then the question must arise of who is to take that decision. It may simply be a matter of professional self-regulation, with published guidelines. But, if there are not sufficient grounds to trust the professionals themselves either to follow the guidelines or to submit themselves to inspection to ensure that they are doing so, then it must be a matter for Parliament, and there will, as I have suggested, need to be legislation. Whether or not, in cases of biotechnology, the professionals are to be trusted will depend on the issues involved. But increasingly, it has to be said, there are huge sums of money to be made by pharmaceutical companies, who may develop their own research teams; increasingly patents are taken out for new techniques, and the competition between companies and consequential secrecy makes any kind of inspection or monitoring nearly impossible. We may therefore see more legislation in the field.

Parliament must obviously be well-informed if it is to produce legislation; for the issues involved will be moral issues, involving public policy of a particular kind, namely, that nothing shall be permitted that is genuinely outrageous to the value that ought to be accorded to human beings. And, as I have said, one cannot make proper moral judgments on a basis of ignorance. Here there is, I believe, a genuine role for a committee of enquiry, or royal commission, composed partly of scientists, partly of practicing doctors, partly of lawyers and perhaps philosophers, or other reasonably level-headed persons, who will make the outcome of their deliberations public, will seek evidence and opinions from as many people as possible before reaching their conclusions, and above all will have the task of educating the general public.

This last point is of the greatest importance. Anonymous departmental civil servants—even if, as one hopes, they are strictly impartial and not under pressure from their ministers—cannot take on the educative role that is necessary in such cases. They cannot write articles or take part in broadcasts or lecture tours to explain the conclusions that they recommend and the arguments on which they are based. Without this fairly lengthy process, no regulatory legislation can be satisfactory. In the field of biotechnology, indeed, it is all too likely that legislation, even if it is not hastily cobbled together to allay public fears, will be over-restrictive and will tend to inhibit valuable research.

There is a difficult balance that must, if possible, be achieved between

CLONING

allowing new technologies to be developed which may have quite unforseen beneficial uses, and on the other hand offending widely held and deep-seated moral feelings. The best that can be hoped for is that by understanding more of the issues in each particular case people may come to feel that the freedoms allowed, and the restrictions imposed, are acceptable, even if not exactly what they would have personally liked to see. It is only if this balance can be achieved that the regulation of technological research can be compatible on the one hand with academic freedom, and on the other hand with a democratic regard for the moral views of people at large.

Cloning
Then and Now

Daniel Callahan

The possibility of human cloning first surfaced in the 1960s, stimulated by the report that a salamander had been cloned. James D. Watson and Joshua Lederberg, distinguished Nobel laureates, speculated that the cloning of human beings might one day be within reach; it was only a matter of time. Bioethics was still at that point in its infancy—indeed, the term "bioethics" was not even widely used then—and cloning immediately caught the eye of a number of those beginning to write in the field. They included Paul Ramsey, Hans Jonas, and Leon Kass. Cloning became one of the symbolic issues of what was, at that time, called "the new biology," a biology that would be dominated by molecular genetics. Over a period of five years or so in the early 1970s a number of articles and book chapters on the ethical issues appeared, discussing cloning in its own right and cloning as a token of the radical genetic possibilities.

Reprinted from *Cambridge Quarterly of Healthcare Ethics* 7, no. 2 (Spring 1998): 141–44. Copyright © 1998 Cambridge University Press. Reprinted with the permission of Cambridge University Press.

CLONING

While here and there a supportive voice could be found for the prospect of human cloning, the overwhelming reaction, professional and lay, was negative. Although there was comparatively little public discussion, my guess is that there would have been as great a sense of repugnance then as there has been recently. And if there had been some kind of government commission to study the subject, it would almost certainly have recommended a ban on any efforts to clone a human being.

Now if my speculation about the situation twenty to twenty-five years ago is correct, one might easily conclude that nothing much has changed. Is not the present debate simply a rerun of the earlier debate, with nothing very new added? In essence that is true. No arguments have been advanced this time that were not anticipated and discussed in the 1970s. As had happened with other problems in bioethics (and with genetic engineering most notably), the speculative discussions prior to important scientific breakthroughs were remarkably prescient. The actuality of biological progress often adds little to what can be imagined in advance.

Yet if it is true that no substantially new arguments have appeared over the past two decades, there are I believe some subtle differences this time. Three of them are worth some comment. In bioethics, there is by far a more favorable response to scientific and technological developments than was then the case. Permissive, quasi-libertarian attitudes toward reproductive rights that were barely noticeable earlier now have far more substance and support. And imagined or projected research benefits have a stronger prima facie claim now, particularly for the relief of infertility.

1. *The response to scientific and technological developments.* Bioethics came to life in the mid- to late-1960s, at a time not only of great technological advances in medicine but also of great social upheaval in many areas of American cultural life. Almost forgotten now as part of the "sixties" phenomenon was a strong antitechnology strain. A common phrase, "the greening of America," caught well some of that spirit, and there were a number of writers as prepared to indict technology for America's failings as they were to indict sexism, racism, and militarism.

While it would be a mistake to see Ramsey, Jonas, and Kass as characteristic sixties thinkers—they would have been appalled at such a label—their thinking about biological and genetic technology was surely compatible with the general suspicion of technology that was then current. In strongly opposing the idea of human cloning, they were not regarded as Luddites, or radicals, nor were they swimming against the tide. In mainline intellectual circles it was acceptable enough to be wary of technology, even to assault it. It is probably no accident that Hans Jonas, who wrote so compellingly on technology and its potentially deleterious effects, was lauded in Germany well into the 1990s, that same contemporary

Germany that has seen the most radical "green" movement and the most open, enduring hostility to genetic technology.

There has been considerable change since the 1960s and 1970s. Biomedical research and technological innovation now encounter little intellectual resistance. Enthusiasm and support are more likely. There is no serious "green" movement in biotechnology here as in Germany. Save possibly for Jeremy Rifkin, there are no regular, much less celebrated, critics of biotechnology. Technology-bashing has gone out of style. The National Institutes of Health, and *particularly* its Human Genome Project, receive constant budget increases, and that at a time of budget cutting more broadly of government programs. The genome project, moreover, has no notable opponents in bioethics—and it would probably have support *even* if it did not lavish so much money on bioethics.

Cloning, in a word, now has behind it a culture far more supportive of biotechnological innovation than was the case in the 1960s and 1970s. Even if human cloning itself has been, for the moment, rejected, animal cloning will go forward. If some *clear* potential benefits can be envisioned for human cloning, the research will find a background culture likely to be welcoming rather than hostile. And if money can be made off of such a development, its chances will be greatly enhanced.

2. *Reproductive rights.* The right to procreate, as a claimed human right, is primarily of post–World War II vintage. It took hold first in the United States with the acceptance of artificial insemination (AID) and was strengthened by a series of court decisions upholding contraception and abortion. The emergence of in vitro fertilization in 1978, widespread surrogate motherhood in the 1980s, and a continuous stream of other technological developments over the past three decades have provided a wide range of techniques to pursue reproductive choice. It is not clear what, if any, limits remain any longer to an exercise of those rights. Consider the progression of a claimed right: from a right to have or not have children as one chooses, to a right to have them any way one can, and then to a right to have the kind of child one wants.

While some have contended that there is no natural right to knowingly procreate a defective or severely handicapped child, there have been no serious moves to legally or otherwise limit such procreation. The right to procreation has, then, slowly become almost a moral absolute. But that was not the case in the early 1970s, when the reproductive rights movement was just getting off the ground. It was the 1973 *Roe* v. *Wade* abortion decision that greatly accelerated it.

While the National Bioethics Advisory Commission ultimately rejected a reproductive rights claim for human cloning, it is important to note that it felt the need to give that viewpoint ample exposure. Moreover, when the commission called for a five-year ban followed by a sunset provision—to allow time for more scientific information to develop and for public discussion to go forward—it surely left the door open for another round of reproductive rights advocacy. For that matter, if the proposed five-year ban is eventually to be lifted because of a change in public attitudes, then it is likely that putative reproductive rights will be a principal reason for that happening. Together with the possibility of more effective relief from infertility (to which I will next turn) it is the most powerful viewpoint waiting in the wings to be successfully deployed. If procreation is, as claimed, purely a private matter, and if it is thought wrong to morally judge the means people choose to have children, or their reasons for having them, then it is hard to see how cloning can long be resisted.

3. *Infertility relief and research possibilities.* The potential benefits of scientific research have long been recognized in the United States, going back to the enthusiasm of Thomas Jefferson in the early years of American history. Biomedical research has in recent years had a particularly privileged status, commanding constant increases in government support even in the face of budget restrictions and cuts. Meanwhile, lay groups supportive of research on one undesirable medical condition or another have proliferated. Together they constitute a powerful advocacy force. The fact that the private sector profits enormously from the fruits of research adds still another potent factor supportive of research.

A practical outcome of all these factors working together is that in the face of ethical objections to some biotechnological aspirations there is no more powerful antidote than the claim of potential scientific and clinical benefits. Whether it be the basic biological knowledge that research can bring, or the direct improvements to health, it is a claim difficult to resist. What seems notably different now from two decades ago is the extent of the imaginative projections of research and clinical benefits from cloning. This is most striking in the area of infertility relief. It is estimated that one in seven people desiring to procreate are infertile for one reason or another. Among the important social causes of infertility are late procreation and the effects of sexually transmitted diseases. The relief of infertility has thus emerged as a major growth area in medicine. And, save for the now-traditional claims that some new line of research may lead to a cure for cancer, no claim seems so powerful as the possibility of curing infertility or otherwise dealing with complex procreation issues.

In its report, the bioethics commission envisioned, through three hypothetical cases, some reasons why people would turn to cloning: to help a couple both of whom are carriers of a lethal recessive gene; to procreate a child with the cells of a deceased husband; and to save the life of a child who needs a bone marrow transplant. What is striking about the offering of them, however, is that it now seems to be considered plausible to take seriously rare cases, as if—because they show how human cloning could benefit some few individuals—that creates reasons to accept it. The commission did not give in to such claims, but it treated them with a seriousness that I doubt would have been present in the 1970s.

Hardly anyone, so far as I can recall, came forward earlier with comparable idiosyncratic scenarios and offered them as serious reasons to support human cloning. But it was also the case in those days that the relief of infertility, and complex procreative problems, simply did not command the kind of attention or have the kind of political and advocacy support now present. It is as if infertility, once accepted as a fact of life, even if a sad one, is now thought to be some enormous menace to personal happiness, to be eradicated by every means possible. It is an odd turn in a world not suffering from underpopulation and in a society where a large number of couples deliberately choose not to have children.

WHAT OF THE FUTURE?

In citing what I take to be three subtle but important shifts in the cultural and medical climate since the 1960s and 1970s, I believe the way has now been opened just enough to increase the likelihood that human cloning will be hard to resist in the future. It is that change also, I suggest, that is responsible for the sunset clause proposed by the bioethics commission. That clause makes no particular sense unless there was on the part of the commission some intuition that both the scientific community and the general public could change their minds in the relatively near future—and that the idea of such a change would not be preposterous, much less unthinkable.

In pointing to the changes in the cultural climate since the 1970s, I do not want to imply any approval. The new romance with technology, the seemingly unlimited aims of the reproductive rights movement, and the obsession with scientific progress generally, and the relief of infertility particularly, are nothing to be proud of. I would like to say it is time to turn back the clock. But since it is the very nature of a progress-driven culture to find such a desire reprehensible, I will suggest instead that we turn the clock forward, skipping the present era, and moving on to one that is more sensible and balanced. It may not be too late to do that.

Further Reading

As you might imagine, the cloning debate has sparked several books, both individually authored and collections by many hands. If you want the scientific background and the events leading up to Dolly, then the place to start is Gina Bari Kolata's *Clone: The Road to Dolly, and the Path Ahead* (New York: W. Morrow, 1998). The author is a skilled science journalist who was indeed the person who first reported the Dolly story (in the *New York Times*). She gives a good account of the science, both the thrill of doing it and the nature of the finished product, and at the same time she has a keen ear for bits and pieces of gossip and scandal. She is not very philosophical but then, why should she be? If this latter is what you are after, then (although more a booklet than a full-length tome) you will find full satisfaction in *The Ethics of Human Cloning*, jointly authored by Leon R. Kass and James Q. Wilson (Washington, D.C.: AEI Press, 1998). You know where Kass stands—he is against it. Wilson, a distinguished political scientist, is no great enthusiast for human cloning, but he does see its worth and feels that in a loving, two-parent family it can be justified. Both authors lay out their positions, and then reply to the other. There is a good sense of the cut and thrust of intellectual debate and, for all that it is a serious subject, there is a liveliness with even an occasional note of humor entering in.

CLONING

An evangelical perspective on cloning is given by Lane P. Lester and James C. Hefley in *Human Cloning: Playing God or Scientific Progress?* (Grand Rapids, Mich.: Fleming H. Revell Co., 1998). They approach the topic with Bible in hand and, as you might expect, worry that cloning is both playing God and scientific progress and that neither alternative is a particularly good thing. Ronald Cole-Turner, who features in our collection, has himself edited a collection on cloning from a religious perspective: *Human Cloning: Religious Responses* (Louisville, Ky.: Westminster John Knox Press, 1997). As in our collection, we find that religious opinion is much divided and not every believer finds the answers as readily as do Lester and Hefley in their book.

More general collections on cloning include *The Human Cloning Debate*, edited by Glenn McGee (Berkeley, Calif.: Berkeley Hills Books, 1998); *Cloning: Science and Society* (Ideas in Conflict Series), edited by Gary E. McCuen (Hudson, Wisc.: G. E. McCuen Publications, 1998); and *Flesh of My Flesh: The Ethics of Cloning Humans: A Reader*, edited by Gregory E. Pence (Lanham, Md.: Rowman and Littlefield, 1998). This final editor has also written a book on the subject, *Who's Afraid of Human Cloning?* (Lanham, Md.: Rowman and Littlefield, 1998), in which he argues that much of the opposition to human cloning is based on outmoded fears of scientific advance, not to mention good, old-fashioned misunderstanding of the facts. Expectedly, Pence does not find many supporters among the conservatives out on the Christian right wing. Nor does Lee M. Silver, author of *Remaking Eden: How Genetic Engineering and Cloning Will Transform the American Family* (New York: Avon, 1998). He is enthusiastic about just about every possible genetic advance and takes cloning to be an obvious tool in the market-economy-fueled march toward happy biological perfection.

Finally let us recommend a good, strong, philosophically informed reader, *Clones and Cloning: Facts and Fantasies about Human Cloning* by Martha C. Nussbaum and Cass R. Sunstein (New York: Norton, 1998). There is material there also on religion and public policy and much more. But before you dash out to buy this and all of the other books listed above, pause and reflect that cloning and indeed all of the genetic technology which surrounds it is within the very fastest moving of scientific-cum-technological areas today. Hence, by all means complement our collection with one or two other works, but then turn your attention to the ongoing items of news and discussions that you find in the leading science journals—more professional publications like *Science* and *Nature* (they both carry excellent general news sections), and the more popular publications like *New Scientist* and *The Sciences*. At the same time, especially if your interest is philosophical and ethical, turn (as we have done) to such periodicals as the *Journal of Medical Ethics* and the *Cambridge Quarterly of Healthcare Ethics*. You will get all of the information and opinion that you need to draw informed judgments of your own.

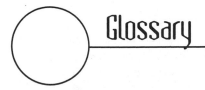

Glossary

A PRIORI: a claim as in logic and pure mathematics, the truth of which is not dependant on experiment or sense experience

AMINO ACIDS: the building blocks of proteins

ANTIBODIES: proteins that bind to foreign chemicals or microorganisms (antigens) to make them harmless; an important component of the immune system

ARTIFICIAL INSEMINATION (AI): a process by which sperm are collected from males and deposited in the female by instruments

ARTIFICIAL SELECTION: selection by humans and breeding of those individual plants or animals that possess desirable traits

ASEXUAL REPRODUCTION: reproduction where there is no fusion of gametes and in which the genetic makeup of parents and offspring is usually identical

BIOETHICS: a discipline concerned with the application of ethics to biological problems, especially in the field of medicine

CLONING

BIOREACTORS: a containment vessel for biological reactions, used in particular for fermentation processes and enzymatic reactions

BIOTECHNOLOGY: technological processes (agriculture, manufacturing, medicine) that involve the use of biological systems

BLASTOCYST: the blastula stage in the development of the mammalian embryo

BLASTOMERE: one cell of the blastula; a cell formed during early cleavage of a vertebrate embryo

BLASTOMERE SEPARATION: the process in which the cells in blastomeres are made to separate; cloning through blastomere separation yields embryos with identical sets of nuclear DNA *and* mitochondrial DNA; also called **embryo splitting**

BLASTULA: usually a spherical structure produced by cleavage of a fertilized egg cell; consists of a single layer of cells surrounding a fluid-filled cavity called the blastocoel

BLOOD PLASMA: the fluid portion of blood containing proteins, salts, and other substances, but excluding the blood cells

CARCINOGEN: a chemical that causes cancer, generally by altering the structure of DNA

CELL CLEAVAGE: first of several cell divisions in early embryonic development that converts the zygote into a multicellular embryo

CELL CULTURE: growth of tissue cells in artificial media

CELL CYCLE: the series of events in the life of a cell consisting of mitosis or nuclear division, cytokenesis or cytoplasmic division, and the interphase during which typically cells grow

CELL DIVISION: the process by which cells multiply

CELLULAR DIFFERENTIATION: see **differentiation**

CHROMATIN: the complex of DNA, protein, and RNA, that makes up chromosomes

CHROMOSOME: a subcellular structure containing DNA plus the proteins that are associated with the DNA

CLONE: multiple copies of identical DNA sequences that are produced when they are inserted into cloning vehicles (plasmids and other vectors) and replicated in bacteria using these cloning vehicles

CLONING: the production of a number of genetically identical DNA molecules by inserting the chosen DNA into a phage or plasmid vector; the vector then being used to infect a suitable host within which the chosen DNA is replicated to form a number of copies (clones)

CONJOINED TWINS: offspring that are physically attached at birth; occurs when the early embryo subdivides forming two groups of cells that do not separate completely; also referred to as **Siamese twins**

CONSEQUENTIALIST ETHICS: morally right actions are those that maximize the good. See **utilitarianism**

CYTOPLASM: general cellular contents exclusive of the nucleus

DEDIFFERENTIATION: the opposite process of differentiation where cells or structures revert back to a less specialized stage

DEONTOLOGICAL ETHICS: a moral theory according to which the rightness or obligatoriness of an action is not exclusively determined by the value of its consequences

DEVELOPMENTAL BIOLOGY: the study of the processes by which cells and cell structures typically become more complex

DIFFERENTIATED CELL: a cell that has undergone the process of cellular differentiation and typically become specialized, e.g., a muscle cell

DIFFERENTIATION: a process changing a young, relatively unspecialized cell to a more specialized cell

DIPLOID: the condition of having a paired set of chromosomes

DIZYGOTIC TWINS: offspring that develop when two eggs are fertilized by a different sperm; also called fraternal twins

DNA (DEOXYRIBONUCLEIC ACID): a long, chainlike molecule that transmits genetic information

E. COLI (ESCHERICHIA COLI): a bacterium whose natural habitat is the gut of humans and other warm-blooded animals

ECTODERM: the outer of the three tissue layers of the early animal embryo; typically gives rise to the skin and nervous systems; see also **mesoderm** and **endoderm**

EMBRYO: a young multicellular organism that develops from the zygote and has not yet become free-living; the developing human organism until the end of the second month, after which it is referred to as a fetus

CLONING

EMBRYO SPLITTING: see **blastomere separation**

EMBRYOLOGY: the scientific study of the development of organisms during the embryonic stage

EMBRYONIC STEM (ES) CELL: a formative cell capable of giving rise to all other kinds of cells

END-IN-ITSELF: a Kantian concept that something has value independent of external material ends

ENDODERM: the inner layer of the three basic tissue layers that appear during early development; becomes the digestive tract and its outgrowths—the liver, lungs, and pancreas; see also **ectoderm** and **mesoderm**

ENUCLEATED CELL OR EGG: a cell or egg where the nucleus has been removed

ENUCLEATION: the process of removing the nucleus from a cell or egg

ENZYME: a protein that speeds or slows a specific chemical reaction

EPIGENESIS: the theory that an embryo develops from a structureless cell by the gradual addition of new parts

EPIGENETIC CHANGE: any change of gene activity during the development of the organism from the fertilized egg to the adult

EPITHELIAL CELLS: cells that make up the tissue that covers body surfaces, lines body cavities, and forms glands

EUGENICS: the claim that the way to improve humankind is through selective breeding

FERTILITY: the state or capacity for producing offspring

FERTILIZATION: the fusion of male and female gametes resulting in the formation of a zygote

FETUS (ADJ. FETAL): an unborn offspring in a late stage of development; from the third month of pregnancy to birth in humans

FIBROBLASTS: connective tissue cells that secrete the intercellular material of, for example, bones and cartilage

GAMETE: a sex cell; in plants and animals, an egg or sperm

GASTRULA: early stage of development that follows the blastula stage

GENE: the unit of heredity that consists of a stretch of DNA; contains information for the construction of other molecules, mostly proteins

GENE THERAPY: the use of DNA sequences either to switch off defective genes or to insert genes that specify the correct product

GENETIC DETERMINISM: the claim that the traits of organisms are a direct function of their genes, not subject to environmental influences

GENETIC DIVERSITY: a measure of the differences in allelic frequencies between isolated or semi-isolated populations due to various evolutionary forces such as selection and genetic drift

GENETIC ENGINEERING: the manipulation of the DNA content of an organism to alter the characteristics of that organism

GENETIC MODIFICATION (GM): a process (especially an artificial process) that results in a change in the genetic makeup of a population

GENETIC SELECTION: artificial selection of specific genes

GENOME: the genetic material of an organism

GENOTYPE: the organism's genetic information, as distinguished from its physical appearance (phenotype)

GERM CELL: cells within the body that give rise to gametes

GESTATION: the duration of pregnancy

G0 PHASE: phase in the mitotic cycle; typically a stage of arrested cellular activity

GROWTH PHASE: see **interphase**

HAPLOID: the condition of having a single set of chromosomes, received from father or mother

HISTOCOMPATIBILITY ANTIGENS: systems of allelic alloantigens that can stimulate an immune response leading to transplant rejection when the donor and recipient are mismatched

HORMONES: "messenger" molecules that coordinate the actions of various tissues of the body; hormones produce a specific effect on the activity of target cells remote from their point of origin

IDENTICAL TWINS: see **monozygotic twins**

CLONING

IMMUNOLOGY: the study of specific defense mechanisms (immune responses)

IMPLANTATION: the attachment of the developing embryo to the uterine wall

INFERTILITY: the inability of a couple to achieve conception (fertilization and establishment of pregnancy) through sexual intercourse

INTERPHASE: phase of the cell cycle during which growth occurs

INTRACYTOPLASMIC SPERM INJECTION: the process by which sperm cells are inserted directly into an egg cell

IN UTERO: occurring in the uterus

IN VITRO: occurring outside a living organism (literally "in glass")

IN VIVO: occurring in a living organism

IVF (IN VITRO FERTILIZATION): the process during which an egg is removed from the ovary and fertilized with sperm in laboratory glassware and then reimplanted in the female

KANTIAN ETHICS: based on the categorical imperative, namely that we should act only according to those maxims that can be consistently willed as a universal law

KERATINOCYTE: a specialized epidermal cell that synthesizes keratin, the substance that makes up nails and feathers

LACTATION: the production or release of milk from the breast

MAMMARY GLAND: the organs that secrete milk to feed offspring

MEIOSIS: process in which a diploid cell undergoes two successive nuclear divisions to produce four haploid cells; the process that produces gametes in animals

MESODERM: the middle layer of the three basic tissue layers that develop in the early embryo; gives rise to connective tissue, muscle, bone, blood vessels, and many other structures; see also **ectoderm** and **endoderm**

METABOLISM: a collective term for the chemical processes involved in the maintenance of life and by which energy is made available

METAPHASE: the stage of mitosis and meiosis during which the chromosomes line up on the equatorial plane of the cell

METAPHYSICS: the branch of philosophy which deals with the ultimate nature and origin of things

MICROINJECTION: injection of cells with solutions by using a micropipette

MISCARRIAGE: the expulsion of the embryo or fetus from the uterus before it is viable

MITOCHONDRIA (SING. MITOCHONDRION): a specialized subcellular structure that converts chemical energy from one form to another

MITOCHONDRIAL DISEASE: diseases that are due to mutations or malfunctions in mitochondria

MITOCHONDRIAL DNA (MTDNA): the DNA contained in the mitochondria

MITOSIS: division of the cell nucleus, resulting in two daughter nuclei with the same number of chromosomes as the parent nucleus; consists of four phases: prophase, metaphase, anaphase, and telophase

MOLECULAR BIOLOGY: the chemistry and physics of the molecules that constitute living things

MONOCLONAL ANTIBODY: an antibody produced by a clone of antibody-producing cells that recognizes only one feature

MONOVULAR TWINS: see **monozygotic twins**

MONOZYGOTIC TWINS: offspring that possess identical sets of genetic material; result from the division of the early embryo; also referred to as identical or monovular twins

MORAL AGENCY: the capability of having rights as an individual

MORPHOLOGY: the scientific study of organic form

MULTIPOTENT: see **totipotent**

MUTAGEN: any agent that is capable of producing changes in the DNA

MUTATIONS: any change in the DNA , including a change in the sequence of nucleotide pairs that constitute DNA or a rearrangement of genes within chromosomes

MUTATIS MUTANDIS: after the necessary or appropriate changes have been made

MYOEPITHELIAL CELLS: epithelial cells of Coelenterata that are provided with basal contractile outgrowths

NATURAL SELECTION: the key Darwinian mechanism for evolutionary change, claiming that only a percentage of organisms in each generation survive and reproduce, and that these do so because they have characteristics which other members of the population do not possess; these "adaptive" characteristics are passed along to offspring

NUCLEAR: pertaining to the nucleus

NUCLEAR DNA: the DNA contained within the nucleus of the cell (as opposed to the DNA contained in the mitochondria)

NUCLEAR SUBSTITUTION: see **nuclear transfer**

NUCLEAR TRANSFER: process where the nucleus of a host cell is removed and replaced with the nucleus of a donor cell that is to be cloned

NUCLEAR TRANSPLANTATION: see **nuclear transfer**

NUCLEOLUS (PL. NUCLEOLI): small nuclear body made up of protein and ribosomal RNA

NUCLEUS (PL. NUCLEI): a double membrane-bounded cellular structure containing DNA

NEURULA: the stage of development of the vertebrate embryo that is characterized by the development of the neural tube from the neural plate

NULLIPOTENT: unable to direct the development of cells unlike themselves

OESTRUS: the recurrent period of heat or sexual receptivity occurring around ovulation in female mammals

ONCOGENESIS: the processes of tumor formation

OOCYTES: cells that give rise to egg cells (ova) by meiosis

OVULATION: the release of a mature egg from the ovary

PARTHENOGENESIS: a form of asexual reproduction in which an unfertilized egg develops into an adult organism

PARTURITION: the process of birth

PATHOGEN: a disease-causing agent

PERINATAL: surrounding the time of birth

PHARMACEUTICALS: chemicals used in medicine

PHENOMENOLOGY: the philosophy in which supposedly we come to know mind as it is in itself through the study of the way in which it appears to us

PHENOTYPE: the physical and behavioral characteristics of an organism

PHYSIOLOGY: the scientific study of the workings of living bodies and their parts

PLOIDY: relating to the number of sets of chromosomes in a cell

PLURIPOTENT: capable of giving rise to various kinds of embryonic tissue

POLYCLONAL ANTIBODY: an antibody produced by many clones of cells that recognize several molecular features of an antigen

POLYMORPHIC LOCI: genetic loci, in a population, which have several different alleles

POST MORTEM: the period after death; also, the examination of an organism after death

PREEMBRYOS: the preimplantation stages of the development of the zygote, occurring during the first two weeks after fertilization

PREFORMATION THEORY: the theory of early developmental biologists that the fully formed animal or plant exists in a minute form in the germ cell (see also **epigenesis**)

PREIMPLANTATION DIAGNOSIS: the determination of the genotype of an in vitro-fertilized human embryo (for genetic diseases) prior to its implantation.

PRENATAL DIAGNOSIS (OR SCREENING): fetal cells are sampled and analyzed for chromosomal and biochemical disorders

PRIMA FACIE: clear and apparent at first glance, presumed true unless disproven by subsequently acquired evidence

PROGENITORS: the ancestors of an organism

PROGENY: offspring

PROMOTER REGION: a site on the DNA that facilitates the initiation of transcription

PRONUCLEUS (PL. PRONUCLEI): one of the two nuclear bodies of a newly fertilized egg cell; the male and the female pronuclei fuse to form the nucleus of the new cell

QUIESCENCE: the state of dormancy or inactivity

RECESSIVE GENE: any gene whose expression is masked by another (dominant) gene

CLONING

RECOMBINANT DNA: hybrid DNA produced by joining pieces of DNA from different sources

REDUCTIO AD ABSURDUM: refutation of an assumption by deriving a contradiction (or otherwise clearly false conclusion) from it

REPRODUCTIVE CLONING: the use of cloning technology to produce new individuals (in contrast to therapeutic cloning)

SENESCENCE: the aging process

SEXUAL REPRODUCTION: form of reproduction in which two gametes fuse to form a zygote

SIAMESE TWINS: see **conjoined twins**

SLIPPERY SLOPE: an argument that an action apparently unobjectionable in itself would set in motion a train of events leading ultimately to an undesirable outcome

SOMATIC CELL: body cell, as distinguished from a germ cell

SOMATIC CELL NUCLEAR TRANSFER (SCNT): the process by which a nucleus from a (somatic) body cell is removed and inserted into an egg cell

SPERMATOGONIUM: germ cell produced by a male

STEM CELL: see **embryonic stem cell**

SURROGATE: a female whose uterus is used to gestate an embryo that is not her own

TELEOLOGY: a theory that describes or explains in terms of purpose

TELOMERASE: the enzyme that elongates telomeres

TELOMERE: the end of a chromosome in which the DNA is looped back

TELOS: endpoint, aim, goal, or purpose of an activity, or a final state of affairs

THERAPEUTIC CLONING: the use of cloning technology for medical purposes, for example, the production of organs for transplantation (in contrast to reproductive cloning)

TISSUE CULTURING: a method of propagating healthy cells outside the organism

TOTIPOTENCY (ADJ. TOTIPOTENT): ability of a cell (or nucleus) to develop into any other type of cell (or nucleus)

TRANSCRIPTION: the process of converting information in DNA into information in RNA (ribonucleic acid); RNA functions mainly in protein synthesis

TRANSGENE: a gene that is taken from one organism and inserted into the genome of another

TRANSGENESIS: genetic modification of animals using transgenes

TRANSGENIC ANIMAL: an animal that has incorporated foreign DNA into its genome

TRANSLATION: the process of converting the information in RNA into protein; also called protein synthesis

TRIMESTER: one of the three three-month periods of human pregnancy

UNDIFFERENTIATED CELL: an unspecialized cell that has not undergone differentiation

UNFERTILIZED EGG: an egg cell that has not undergone fertilization

UTILITARIANISM: the moral theory that an action is morally right if and only if it produces at least as much good (utility) for all people affected by the action as any alternative action

VECTOR: an agent, such as a bacterial plasmid or virus, that transfers genetic information

XENOTRANSPLANTATION: transplantation of tissue or organs between species

ZYGOTE: one-celled embryo formed by the fusion of two germ cells

 # Contributors

JUSTINE BURLEY has recently edited two books: *The Genetic Revolution and Human Rights* (The Oxford Amnesty Lectures, 1998) (Oxford University Press, 1999) and *Dworkin and His Critics* (Basil Blackwell, 1998).

D. CALLAHAN has a Ph.D. in philosophy from Harvard. He was a cofounder of the Hastings Center in 1969, and was the director and president until 1996. He is an elected member of the Institute of Medicine, National Academy of Sciences; a member of the Director's Advisory Committee, Centers for Disease Control and Prevention, and chair of its ethics committee. He is also the 1996 winner of the Freedom and Scientific Responsibility Award of the American Association for the Advancement of Science.

KEITH H. S. CAMPBELL was Ian Wilmut's cocreator of Dolly. He is a cell-cycle biologist that discovered how to sync cells at stage "G0" in order to attempt cloning. His work was essential for the breakthrough that lead to Dolly and future cloning.

CLAIRE COCKCROFT writes for the *Guardian*.

CLONING

RONALD COLE-TURNER is the H. Parker Sharp Professor of Theology and Ethics at Pittsburgh Theological Seminary, a position which relates theology and ethics to developments in science and technology. He is an ordained minister of the United Church of Christ and chairs the UCC committee on genetics. He is also the author of the popular baptism hymn, "Child of Blessing, Child of Promise."

DAVID CONCAR is a freelance writer for *New Scientist* and the *Guardian*. He writes on issues concerning cloning, genetically modified foods, and drug use, especially Ecstasy.

DOUGLAS COUPLAND is the author of several books: *Generation X*, *Shampoo Planet*, *Life After God*, *Microserfs*, and *Girlfriend in a Coma*.

DAVID ELLIOTT is assistant professor in the Department of Philosophy and Classics at the University of Regina. His main research interests are in ethical theory, applied ethics (particularly biomedical ethics), philosophy of law, and political philosophy.

WALTER GLANNON was a fellow at the Institute for Ethics at the American Medical Association from September 1998 to June 1999. He has a Ph.D. in Spanish Literature from Johns Hopkins University and a Ph.D. in Philosophy from Yale. Dr. Glannon has taught literature, philosophy, and biomedical ethics at Smith College, Simon Fraser University, the University of Calgary, and McGill University. His research interests in biomedical ethics include death and dying, allocation of scarce medical resources, and genetics.

STEPHEN JAY GOULD is Agassiz Professor of Zoology at Harvard University.

JOHN B. GURDON is Master of Magdalene College, Cambridge, and a Governor of the Wellcome Trust. Since 1991, he has been the chairman of the Wellcome/CRC Institute for Cancer and Developmental Biology, Cambridge.

JOHN HARRIS is professor of bioethics and applied philosophy and research director at the Centre for Social Ethics and Policy at the University of Manchester. His main research/teaching interests are the ethics and social policy implications of HIV/AIDS, the ethics and social policy implications of communicable diseases more generally, the ethics of tolerable risk and public health, the implications of the new reproductive technologies, children's rights, discrimination, the ethics of the treatment of the old, and those with short life expectancy.

SØREN HOLM is reader in bioethics at the Institute of Medicine, Law, and Bioethics (a collaboration between the Universities of Liverpool and Manchester with support from the North West Regional Health Authority) and the Centre for Social Ethics and Policy at the University of Manchester. He is honorary registrar in the Department of Medical Oncology, Christie Hospital, where he works in the ovarian cancer clinic. He is a member of the Danish Council of Ethics, and has chaired working groups of the council on a number of issues including Resource Allocation in the Danish Health Care System, Compulsory Treatment and Care in Psychiatry, The Conditions of the Demented Elderly, and Transplantation issues.

AXEL KAHN is director of the INSERM laboratory of Research on Genetics and Molecular Pathology at the Cochin Institute of Molecular Genetics, 75014 Paris France.

LEON R. KASS is the Addie Clark Harding Professor in the Committee on Social Thought and the College at the University of Chicago.

A.J. KIND works for PPL Therapeutics Ltd., one of the world's leading companies in the application of transgenic technology to the production of human proteins for therapeutic and nutritional use.

PHILIP KITCHER is professor of philosophy at Columbia University. He is author of *The Lives to Come: The Genetic Revolution and Human Possibilities.*

ARLENE JUDITH KLOTZKO, a lawyer and bioethicist based in New York City, has edited an anthology on cloning for Oxford University Press.

RONALD A. LINDSAY holds a law degree from the University of Virginia and a Ph.D. in philosophy from Georgetown University. He writes on bioethics issues.

J. H. LIPSCHUTZ is in the department of Anatomy & Medicine at the University of California San Francisco.

J. MCWHIR researches control of the establishment of cell lineage, developing murine embryo stem cells, and gene targeting at the Roslin Institute in Scotland.

HARRY M. MEADE is vice president of transgenic research at Genzyme Transgenics Corporation in Framingham, Massachusetts. He has published papers on transgenic milk production in *Biotechnology* and other journals, and has authored several relevant patents. Dr. Meade also is a coinventor on the first issued patent relating to the production of therapeutic proteins in the milk of animals.

BERNARD E. ROLLIN is professor of philosophy at Colorado State University. He edited the two-volume *The Experimental Animal in Biomedical Research* (1989, 1995). He is one of the leading scholars in animal rights and animal consciousness and is a weight-lifter, horseman, and motorcyclist.

JULIAN SAVULESCU is a research fellow and group leader of the ethics unit of the Murdoch Children's Research Institute, Melbourne.

A. E. SCHNIEKE works for PPL Therapeutics Ltd., one of the world's leading companies in the application of transgenic technology to the production of human proteins for therapeutic and nutritional use.

MARY WARNOCK, a life peer, is the former mistress of Girton College, Cambridge and honorary fellow of Lady Margaret Hall and St. Hughs College, Oxford. During her career as a philosopher she chaired a number of significant government inquiries, including those on Animal Experimentation, Special Education, Environmental Pollution, and Human Fertility and Embryology.

ROBERT WILLIAMSON, FRS, is scientific director of the Murdoch Children's Research Institute, a globally competitive research institute with a unique focus on the health of babies, children, and adolescents.

IAN WILMUT researches early mammalian development, embryo manipulation, nuclear transfer, and gene targeting in mice, cattle, sheep, and pigs at the Roslin Institute, a nonprofit organization sponsored by the Biotechnology and Biological Sciences Research Council (BBSRC) which is the UK's leading funding agency for academic research in the nonmedical life sciences and is one of seven research councils established by Royal Charter in 1994.

ROBERT WINSTON, professor of infertility medicine, Royal Postgraduate Medical School, Hammersmith Hospital, London, is an in vitro fertilization pioneer. He developed gynecological microsurgery for the treatment of tubal disorders in the 1970s and better techniques for sterilization reversal. He was scientific adviser on contraceptive matters and reproductive research to WHO from 1975 to 1980 and has published approximately three hundred scientific publications in learned journals.